Radioactive Waste
Disposal and Geology

Topics in the Earth Sciences

SERIES EDITORS

T.H. van Andel
Stanford University

Peter J. Smith
The Open University

Radioactive Waste Disposal and Geology

Konrad B. Krauskopf
Stanford University

London   New York
CHAPMAN AND HALL

First published in 1988 by Chapman and Hall Ltd
11 New Fetter Lane, London EC4P 4EE
Published in the USA by Chapman and Hall
29 West 35th Street, New York NY 10001

© *1988 Krauskopf*

Printed in Great Britain at the
University Press, Cambridge

ISBN 0 412 286300 (HB)
ISBN 0 412 286408 (PB)

British Library Cataloguing in Publication Data

Krauskopf, Konrad
  Radioactive waste disposal and geology.—
  (Topics in earth sciences).
  1. Radioactive waste disposal
  I. Title   II. Series
  621.48'38   TD898

  ISBN 0-412-28630-0
  ISBN 0-412-28640-8 Pbk

Library of Congress Cataloging in Publication Data

Krauskopf, Konrad Bates, 1910–
  Radioactive waste disposal and geology / Konrad B. Krauskopf.
    p.   cm. — (Topics in the earth sciences)
  Bibliography: p.
  Includes index.
  ISBN 0-412-28630-0.   ISBN 0-412-28640-8
  (pbk.)
  1. Radioactive waste disposal.   2. Engineering geology.
  I. Title.   II. Series.
  TD898.K73   1988
  621.48'—dc19

Contents

Preface

The perception of radioactive waste as a major problem for the industrial world has developed only recently. Four decades ago the disposal of such waste was regarded as a relatively minor matter. Those were the heady days when nuclear fission seemed the answer to the world's energy needs: the two wartime bombs had demonstrated its awesome power, and now it was to be harnessed for the production of electricity, the excavation of canals, even the running of cars and airplanes. In all applications of fission some waste containing radioactive elements would be generated of course, but it seemed only a trivial annoyance, a problem whose solution could be deferred until the more exciting challenges of constructing reactors and devising more efficient weapons had been mastered. So waste accumulated, some in tanks and some buried in shallow trenches. These were recognized as only temporary, makeshift measures, because it was known that the debris would be hazardous to its surroundings for many thousands of years and hence that more permanent disposal would someday be needed. The difficulty of accomplishing this more lasting disposal only gradually became apparent. The difficulty has been compounded by uncertainty about the physiological effects of low-level radiation, by the inadequacy of detailed knowledge about the behavior of engineered and geologic materials over long periods under unusual conditions, and by the sensitization of popular fears about radiation in all its forms following widely publicized reactor accidents and leaks from waste storage sites.

So in the present world the management of nuclear waste looms as a major dilemma for many countries. How serious a problem is it? Is radioactive waste really an intractable menace to this and all future generations, or are we wasting billions of dollars on what is essentially a trivial matter? Answers to this question cover a wide spectrum, ranging from calm assurance that disposal could be accomplished easily within a decade to apocalyptic fears that the problem is essentially insoluble. Certainly the amount of radioactive waste is trivial compared with the huge volumes of other toxic industrial waste, but just as certainly its radioactive content gives it properties that complicate its management enormously. Even among experts there is no agreement as to how much effort is warranted to ensure the safe disposal of various kinds of nuclear waste.

Putting the waste underground is the generally accepted long-term solution for the disposal problem – deep underground for the long-lived, highly radioactive waste, in shallow trenches for low-level material – but details of where, when, and how this is to be accomplished continue to excite lively debate. Geologists are called on for recommendations, since the subsurface is the province of their supposed expertise, but making predictions about the long-term behavior of hot radioactive material buried in rock is a problem beyond their normal experience. Even those who have studied the matter carefully have differences of opinion about the safety of disposal in various environments, and geologists unacquainted with the problem share the wide divergence of sentiment that is prevalent in the general public. It is to this latter group – readers with some geologic background but no professional concern with the peculiarities of radioactive waste – that this book is primarily addressed. My hope is to summarize the main features of the disposal problem in sufficient depth that earth scientists of all stripes can understand the background of current arguments, without being burdened by tedious detail.

The book should also be useful to a wider audience. Although it is designed for readers with some technical training, its geologic explanations are intended to supply depth for those interested but not to constitute the main thread of the story. The problem should be of general interest, not only in itself but because it exemplifies a kind of dilemma that is all too common in the modern world: a problem involving both technology and politics, about which strong emotions are aroused and on which technical experts disagree, and which will ultimately require a political decision based on necessarily imperfect knowledge.

I have tried to present a world-wide overview of current thinking about disposal questions, but because my experience is chiefly in the United States there is inevitably a preponderance of American examples and expressions of opinion. I owe a debt of thanks to my many friends and colleagues, American and Canadian and European, with whom I have worked and argued over the years in committees and symposia and on field trips, for the education on waste management questions that has made this book possible. In addition, for much help in clarifying ideas and spotting infelicities of expression, both my readers and I are indebted to the patience and persistence and good sense of my wife.

Konrad B. Krauskopf
1987

Foreword

During the past four decades, the diversity within the Earth sciences has increased enormously – in depth, in breadth, in technology, and in the level of sophistication of the use of physics, chemistry, biology, and the space sciences – with an inevitable increase in the degree to which rampant specialization has isolated the practitioners of many subfields. Ironically, the potential value of various specialities for other, often non-contiguous, ones has also increased. Not so very long ago, for example, the findings of marine geology were of interest to oceanographers alone. Today they are crucial in the understanding of the genesis of the continents as well. What is at the present time quiet, unseen work in a remote corner of our discipline, may tomorrow enhance, even revitalize some quite different subject.

Inevitably, but also most regrettably, communication between subfields has become much more difficult, for a variety of reasons. While the number of research reports published annually keeps rising exponentially, the reading time available to individual scientists has, if anything, decreased. Furthermore, many subfields now draw to such a degree on sophisticated methodology and on fundamentals taken from other disciplines, not to mention a level of mathematics opaque to the majority of Earth scientists, that the understanding of much writing often remains limited to immediate peers. It may well be that the literature on the deep seismic structure of the continental crust is not readily accessible to the sedimentologist, but that does not mean that it would not at some point be very helpful, or more generally be simply of interest. The enormous proliferation of specialist journals does not help, and there is reason to doubt that the expanded volume of printed matter corresponds to an equal increase in information. There are, of course, review papers and review journals, but they, too, tend to be designed for specialists and to possess a high access threshold for outsiders. Moreover, the areas reviewed have themselves narrowed to the point where one at times wishes for reviews of review papers.

This, most would agree, is not just unfortunate but, if accepted with a sigh of resignation, dangerous. It is dangerous to limit, for lack of mutual comprehension, the flow of information in the Earth sciences from one subfield to another. It is even more dangerous to make it difficult for students of different subfields even to know what their colleagues are up to. Certainly, one's awareness of one's need to know may inspire a search and that search, though usually tedious, is likely to be successful. But what if one does not even know that relevant work is being done in some other part of the Earth sciences?

We regard it as even more important that most of us, specialists that we may have become, began our careers because of a deeply felt curiosity for the Earth as a whole; and we remain at heart, although often helplessly, reluctant to give this broad perspective up for a narrow view of only the continental margin, high temperature high-pressure geochemistry, or whatever. Apart from the obvious utility of a broader perspective, it also guarantees periodic renewal of one's drive and enthusiasm.

Some years ago we began to think about a series of concise, modestly demanding books that would, for Earth scientists across the board, illuminate the state of the art in as many subfields as possible in a language we would all have in common. The idea was not immediately welcomed with great warmth by either potential publishers or authors. Such books, each envisaged for reading at leisure, written in a style and language calculated to hold one's interest, and referenced only to a level that permits more advanced pursuit, would fall in an uncomfortable grey zone between undergraduate textbook and specialty monograph for which the market has rarely been tested. Moreover, the requirement that they would be thorough but easily read, captivating yet balanced, and personal though fair, caused us to wonder whether enough authors could be found with the required skills and, more importantly, the time and will to try their hand at what is, for most of us, a new medium of expression.

The appearance in print of this Foreword indicates that we did indeed find a publisher willing to experiment with the idea, and persuaded what is now a growing number of authors to make it a reality. For their encouragement and that of Chapman and Hall, and especially for the enthusiasm of our first editor, Dr. Alan Crowden, we are most grateful.

This, then, is one book in what we hope will be a long and diverse series documenting the current state of the art in many fields cultivated by the Earth sciences today. We have for this series no orderly plan; the diversity of subject matter, our wish to emphasize fields of current interest, and the real difficulty of finding wise and experienced authors with consummate writing skills, prevent such a systematic approach. How long and how diverse the series will ultimately become depends on this pool of authors and, above all, on the response among our colleagues.

Tjeerd H. van Andel
University of Cambridge

Peter J. Smith
The Open University

1

Introduction

1.1 INDUSTRIAL WASTE AND RADIOACTIVE WASTE

Industrial waste is a major nuisance, and its disposal has become a serious problem. The problem concerns geologists because a common disposal method is to put the waste on or under the ground surface, where it is exposed to water that can dissolve some of its toxic ingredients. The dissolved material may then be carried by surface runoff or underground flow to a point where the water is used for domestic or recreational purposes. To prevent or to minimize the dissolution and transport of toxic substances requires that the geology of the disposal site be meticulously studied and the disposal procedure well engineered. Places for disposal are sought where movement of surface water can be controlled, where the amount of underground water is small and its movement slow, and where the rock and soil through which the water moves can be trusted to prevent, or at least to retard, the migration of its hazardous solutes.

How much retardation is needed depends on the kind of toxic material the waste contains, and particularly on how long the toxicity will persist. Some organic compounds decay within a few years or decades, and hence need only be moderately delayed in their travels with ground water. The more resistant compounds and poisonous metals that some industrial waste contains remain hazardous practically forever, and they must be prevented from escaping in more than minute traces far into the future. The geology of an intended disposal site must therefore be suited to the nature of the waste that will be placed in it.

If the waste contains radioactive material, its disposal becomes peculiarly difficult. Concentrations of radioactive elements in water escaping from a disposal site must be kept absurdly low, because such elements, if taken into the body by ingestion or inhalation, are a potential cause of radiation damage to cells even in extremely small amounts. Waste with a high level of radioactivity gives off so much radiation that it cannot be handled directly, but requires remote control equipment. High-level waste is also thermally hot, so that if put underground it will subject the surrounding rock to both heat and intense radiation for a long time. In engineering experience we have no precedent for such a situation, and its geologic consequences are hard to predict. Will high-level radioactive waste remain safely isolated underground, or will the heat and radiation it generates so disturb the enclosing rock that eventually some radioactive atoms will escape? It is small wonder that experts disagree about the answer to this question, because past experience gives us no guidance. For other kinds of industrial waste we have many decades of past disposal efforts as background, but with radioactive waste we embark on an undertaking of which many aspects are wholly new.

Radioactive elements are continually decaying, at a rate that is specific for each kind of radioactive atom. If we could guarantee the isolation of waste containing such elements for a long enough period, the radioactivity would largely disappear and the waste would become harmless (except for possible chemical toxicity). For some elements the period is short, a matter of a few years or decades, and if a batch of waste contains only these elements the problem of disposal is greatly simplified. But other radioactive elements decay only slowly, and waste containing them remains hazardous for times extending to tens or hundreds of thousands of years. It is the need to guarantee isolation from the human environment for these enormous lengths of time that makes the disposal of radioactive waste so difficult.

Is geologic knowledge sufficient to give confidence in such a guarantee? Much debate has been aroused by this question. Some have felt that uncertainties about geologic disposal are so great that a more radical solution to the handling of radioactive waste must be sought – say, transmuting the elements to reduce their radioactivity, or sending the waste in rockets to outer space. Several such alternatives have been explored, but prevailing opinion has come back to geologic disposal as the safest and most economically reasonable method of managing waste. The argument is not yet dead, but in all countries that have given serious attention to the disposal problem, burial of most radioactive waste underground is the procedure favored by a majority of those technically competent to judge. For high-level waste, however, subsurface disposal has not yet won wide political acceptance.

Disposing of waste anywhere under the land surface means that the waste will be located at no great distance from habitations, and a negative reaction from the local populace is practically assured. Persuading communities to countenance the thought of an underground repository nearby is still an unresolved part of the waste management problem. Such persuasion will require, among other things, a wider appreciation of the role geology can play in ensuring the safety of an underground disposal site.

In this book we set out to explore various aspects of the disposal problem, with emphasis on the strengths and weaknesses of the proposal to solve the problem on a largely geological basis.

1.2 RADIATION

We start with some generalities about the properties of radiation and radioactive substances.

Radioactive elements, by definition, continually emit radiation in the form of small particles, with or without accompanying emission of electromagnetic waves similar to X-rays. To such radiation our normal senses do not respond; a piece of high-grade uranium ore for example, is not distinguished from other kinds of heavy black rock by anything that can be seen, felt, or smelled. The radiation is readily detected however, with a Geiger–Müller counter or similar instrument, or by darkening of a photographic plate on which the ore specimen is placed. The response of instrument or photographic plate is due to the ability of the radiation to ionize (displace electrons in) atoms which it strikes; because of this ability the radiation is described as *ionizing radiation*, in contrast to non–ionizing radiation such as visible light. A further effect of radioactivity is a rise in temperature, simply because the energy of the radiation is converted into heat as it moves through matter. The temperature increase is too minute for detection in uranium ore, but would be readily apparent in a piece of high-level radioactive waste. To feel the heat by handling a piece of waste is hardly advisable, however, because such concentrated radiation is destructive to living flesh.

The damaging effect of radiation is a result of the ionization of atoms through which it moves: if the atoms are part of living cells, the compounds to which they belong are broken down and the cells cease to function. When many cells are so affected, sickness or death can result. Even if no immediate effect is noticeable – if the radiation is of only moderate intensity, or if exposure is brief – enough cells may still be damaged to cause cancer to develop several years later, or cells of reproductive organs may be altered sufficiently to cause genetic damage in offspring. The physiological effects of radiation are potentially so drastic and so mysterious that radioactive waste in all its forms strikes terror in the hearts of many people. The terror is partly irrational, because with care radioactive substances can be safely handled, but it is sufficiently justified to warrant extreme caution in any procedure for disposing of waste containing such materials.

Radiation damage to living organisms depends in a complex way on the intensity of the radiation, the kind of radiation, and the length and continuity of exposure. It would be convenient if we could specify just what intensity and what exposure times are needed to make radiation hazardous, but unhappily this turns out to be difficult. If the intensity is low enough, exposure over a whole lifetime is not sufficient to cause damage: this is evident from the fact that we are exposed to low-level ionizing radiation every minute of our lives. Rocks and soil contain uranium and thorium, cosmic rays from outer space produce radioactive atoms as they travel through the atmosphere, the food we eat contains radioactive potassium and carbon, X-rays used in medical diagnosis and therapy are ionizing radiation similar to that from radioactive atoms: we couldn't conceivably escape from the low-level radiation in the world around us even if we wanted to. We give no thought to this natural radiation, and rightly so. From all the evidence we have, it does us no harm. It might even be good for us: like many substances that are poisonous in large amounts but essential to life in traces (chromium and selenium are examples), radiation in minute doses might be necessary for our existence. There is no evidence for such a speculation, but there is similar lack of evidence for any harm from natural levels of radiation.

The difficulty in setting a threshold for ionizing radiation – a level below which the radiation does no damage, and above which it becomes hazardous – is in large part the familiar difficulty of statistical 'noise'. How could such a threshold be established? The obvious experimental attack (if the welfare of our experimental subjects were not a concern) would be to start with natural radiation, which we assume to be harmless, and gradually increase the amount until physiological damage to a considerable population is detectable. In a sense a part of the experiment is done for us, because natural radiation varies in both space and time. Cosmic rays for example, are attenuated as they pass through the atmosphere, so that radiation levels increase with altitude; citizens of Denver are exposed to about twice as much natural radiation as those who live in New Orleans, just because of the elevation difference. If the threshold is low, this difference should lead to a difference in health characteristics between the two cities, say a difference in the incidence of cancer. But such a difference would be impossible to detect, because cancer incidence depends on so many other attributes of the two populations – mean age, diet, smoking habits, occupations, and so on. If the dose to Denverites were increased artificially, by way of experiment, it would have to rise far above the natural level before any effect could with certainty be ascribed to radiation rather than to other factors in the statistical noise; and even if an effect were seen, there would still be argument as to just where the threshold should be placed.

Experiments with animals can be better controlled, but still the effects of very low radiation levels become lost in the statistical background. Some individual animals are more resistant to radiation than others, and the kind of damage produced by radiation is similar to damage caused naturally by chance mutations or accidents in the animals' lives; sorting out one from the other is impossible. At higher radiation levels the effects on animals are roughly proportional to radiation dose, and one way to guess at the hazard of low doses is simply to extrapolate the proportionality downward. This means we assume that no threshold exists, that exposure to radiation is potentially harmful at any level. In other words, even natural radiation is assumed to be damaging, and most of us survive only because the effects of radiation build up so slowly that we die from other causes before they become apparent.

This assumption, that the proportionality between dose and physiological effect holds even for the lowest doses, is called the 'direct proportion, no threshold hypothesis'. It is extremely conservative in the sense that it may well exaggerate the hazard from weak radiation, but it is nevertheless generally adopted by regulatory agencies in setting standards for permissible radiation levels. For such standards, it is certainly preferable to err on the conservative rather than the liberal side. The assumption however, gives little guidance as to what the standards should be; if any radiation is potentially harmful, the only safe standard is no radiation at all. This is manifestly silly, because we can't escape the sea of natural radiation in which we live. The only reasonable basis for standards then, is a comparison with this natural radiation. If additions to radiation dose can be kept within, or close to, the observed variations in the natural level, we should have little cause for worry. Temporary increases are permissible, but cumulative doses over periods of months or years should not be allowed to greatly exceed natural levels. Reasoning of this sort lies behind regulatory limits set for permissible concentrations of radioactive material in food and water and the amounts that can be allowed to escape from nuclear reactors and waste-disposal sites.

Numbers are useful to put radiation doses in perspective. The basic unit of absorbed dose of ionizing radiation is the rad, defined as the dose resulting from absorption of 10^{-5} joules of radiation energy per gram of absorbing material. Because the biological effects of different kinds of radiation are not the same, a more convenient unit of dose is one adjusted for these different kinds, which permits direct comparison of the biological effects of different radiation sources. This unit is the rem, equivalent to the rad for some kinds of radiation but somewhat modified for others. In the Système Internationale, the preferred unit is the sievert, equal to 100 rem. A dose of 0.5 Sv to a human being in the course of a day results in radiation sickness within a few hours or days. From a short-term dose of 10 Sv, death is nearly certain; from 4 Sv, the chance of death is about 50%. Long-term exposure to lower levels, say of the order of a few hundredths of a sievert per year, can result in general ill health and the possible eventual development of cancer or genetic defects. Natural radiation gives an average annual dose in most places of a little over 0.001 Sv (1 millisievert); added to this is about a third more from X-rays used for medical purposes. The natural background radiation varies from place to place, primarily because of greater exposure to cosmic rays at high altitudes; in the United States it ranges from about 0.9 millisieverts per year in Louisiana to 1.8 in Colorado. Because the cosmic ray flux reaching the earth is variable, natural radiation at any one place varies with time over a range of a few tenths of a millisievert. The differences in natural levels are not generally perceived as having an influence on daily life – one doesn't avoid a trip to Denver, or a flight in a high-altitude jet, because of greater exposure to radiation – and the differences provide a basis, as noted above, for decisions by regulators about what increases in radiation due to waste-disposal operations will be acceptable.

1.3 RADIOACTIVITY

Radioactivity, the spontaneous emission of radiation, is a property of a few naturally-occurring elements of high atomic mass and some minor isotopes of a few lighter elements. Uranium and thorium are examples of the heavy elements, ^{40}K and ^{14}C of the light isotopes. (Numbers in such symbols refer to atomic masses, roughly the ratio of each atomic mass to that of hydrogen.) Radioactive isotopes of many other elements can be prepared artificially by bombardment with fast-moving particles.

Emission of radiation from any radioactive isotope means the loss of small particles from the nuclei of its atoms. By losing particles, the nuclei are transformed into nuclei of a different element. This second element may also be radioactive, its atoms decaying to yet another kind, the sequence continuing until ultimately a nucleus is formed that is stable. Table 1.1 is an example of such a disintegration sequence, starting with ^{238}U and ending with stable ^{206}Pb. The radioactive atoms of whatever sort are spoken of as *radionuclides*, and the atoms produced by radioactive decay are called *daughter atoms*, or atoms of a *daughter element*.

Radioactive nuclei decay individually, the number of disintegrations per second being a measure of the amount of radioactive material present. The common unit of this amount is the curie, defined as a quantity of radioactive material in which 3.7×10^{10} (37 thousand million) atoms disintegrate each second. This is approximately the number of disintegrations taking place per second in one gram of the element radium. In the Système Internationale, the preferred unit is the becquerel (Bq), equal to 1 disintegration per second. The amount of radioactivity in a source is obviously related to the resulting radiation dose to objects in its vicinity,

Table 1.1 Example of a radioactive decay series. Three other long decay series are known: $^{235}U \rightarrow ^{207}Pb$, $^{232}Th \rightarrow ^{208}Pb$, $^{237}Np \rightarrow ^{209}Bi$

Radioactive isotopes	Kind of radiation	Half-lives
^{238}U ↓	alpha, gamma	4.5×10^9 yr
^{234}Th ↓	beta, gamma	24 da
^{234}Pa ↓	beta, gamma	6.8 hr
^{234}U ↓	alpha, gamma	2.5×10^5 yr
^{230}Th ↓	alpha, gamma	80 000 yr
^{226}Ra ↓	alpha, gamma	1 600 yr
^{222}Rn ↓	alpha	3.8 da
Four isotopes with half-lives of less than 30 min ↓		
^{210}Pb ↓	alpha, gamma	22 yr
^{210}Bi ↓	beta	5 da
^{210}Po ↓	alpha, gamma	138 da
^{206}Pb		stable

but the relation is not simple because the dose varies with the kind of disintegrating atom and, of course, with distance from the source.

The proportionate amount of a radioactive isotope that decays in a given time period is constant. Thus if half of a 1 gram mass decays in 10 years, half of the remainder (0.25 g) will decay in the next 10 years, 0.125 g in the next 10 years, and so on. The time required for half of any given mass to decay is called the *half-life* of the isotope.

Half-lives have an enormous range, from fractions of a second to many millions of years, as is illustrated in Table 1.2. The table includes isotopes mentioned in the text and a few others often referred to in discussions of waste disposal. The entry for each isotope includes the common name, the chemical symbol, the kind of particle emitted, and the half-life. The numbers in the symbols are the weights or masses of the atoms, expressed as (roughly) the number of times heavier a given atom is than an atom of

Table 1.2 Glossary of radioisotopes

Isotope	Symbol	Emitted particle	Half-life
americium-241	^{241}Am	alpha	433 yr
americium-243	^{243}Am	alpha	7 370 yr
bismuth-210	^{210}Bi	beta	5.0 da
radiocarbon	^{14}C	beta	5 730 yr
curium-245	^{245}Cm	alpha	8 500 yr
cobalt-60	^{60}Co	beta	5.27 yr
cesium-135	^{135}Cs	beta	3×10^6 yr
cesium-137	^{137}Cs	beta	30.17 yr
tritium	^{3}H	beta	12.33 yr
iodine-129	^{129}I	beta	1.6×10^7 yr
krypton-85	^{85}Kr	beta	10.7 yr
neptunium-237	^{237}Np	alpha	2.14×10^6 yr
protoactinium-234	^{234}Pa	beta	6.8 hr
lead-210	^{210}Pb	beta	22.3 yr
polonium-210	^{210}Po	alpha	138.4 da
plutonium-239	^{239}Pu	alpha	24 000 yr
plutonium-240	^{240}Pu	alpha	6 570 yr
radium-226	^{226}Ra	alpha	1 630 yr
radon-222	^{222}Rn	alpha	3.82 da
tin-126	^{126}Sn	beta	ca. 10^5 yr
strontium-90	^{90}Sr	beta	28.8 yr
technetium-99	^{99}Tc	beta	2.14×10^5 yr
thorium-230	^{230}Th	beta	80 000 yr
uranium-234	^{234}U	alpha	2.45×10^5 yr
uranium-235	^{235}U	alpha	7.04×10^8 yr
uranium-238	^{238}U	alpha	4.47×10^9 yr
xenon-133	^{133}Xe	beta	5.25 da

hydrogen. (Strictly, the standard of comparison is 1/12 of the mass of an atom of carbon-12, rather than an atom of hydrogen. Alternatively and more precisely, the numbers are *mass numbers*, equal to the sum of the numbers of protons and neutrons in the nuclei of the atoms.)

The half-life of an isotope is a rough inverse measure of the intensity of the radiation it generates, since the greater number of disintegrations in a shortlived material clearly must produce stronger radiation. Half-lives are important also as an indication of how long the radiation from a given isotope will remain potentially hazardous: by way of a very rough rule-of-thumb, in most circumstances the amount of an isotope remaining after 10 half-lives is so small that its radioactivity is no longer a serious threat to the surroundings.

Two kinds of particles make up the bulk of those emitted from radioactive substances: nuclei of helium atoms called *alpha particles* or *alpha rays*, which consist of two protons and two neutrons, and electrons, called *beta particles*. The electromagnetic waves that may accompany (or closely follow) emission of either alpha or beta particles are called *gamma rays*. The gamma rays, like X-rays but with shorter wavelengths, are the most penetrating of the three kinds of radiation, traveling many metres through air before their energy is dissipated and requiring at least a few centimetres of a heavy metal like lead for effective shielding. Because gamma rays readily penetrate the body, they are the most damaging of the three from an external source like radioactive waste, and the kind that makes necessary substantial shielding and the handling of high-level waste by remote control. Beta rays are absorbed by a few meters of air and by small thicknesses of metal or glass. If skin is unprotected they can penetrate flesh sufficiently to cause burns, and can do serious harm if a beta-ray source is taken into the body. Alpha rays, heaviest and slowest of the three, are stopped by a sheet of paper or by the outer layer of skin. They do no damage from an external source, but if even traces of an alpha-emitting isotope enter the body by ingestion or inhalation the heavy particles can cause serious disruption of nearby cells. Emission of alpha particles is largely confined to heavy elements, i.e. to radioactive isotopes with atomic masses greater than that of bismuth (209); isotopes emitting beta particles are known for all elements, from the heaviest to the lightest.

Some radioactive waste contains only a single isotope, for example discarded containers of ^{60}Co, an isotope extensively used in radiotherapy. Most waste however, and certainly the great bulk of the high-level waste that poses the most difficult problem of disposal, is made up of many different isotopes with a wide range of half-lives and a great variety of chemical properties. Some waste consists mostly of beta-emitting nuclides, some mostly of alpha-emitters, some a mixture of the two. The activity of most high-level waste decreases rapidly at first as the short-lived isotopes decay, but many remain at a sufficient level to be hazardous for very long times if long-lived isotopes are present.

Because radioactivity is a property of the atomic nucleus rather than the outer part of the atom, it is not affected by chemical reactions or by the temperatures and pressures reached in ordinary laboratory or industrial work. A common query when laymen ponder the problems of waste disposal is, 'Why not just incinerate the stuff, or neutralize it, the way you would a troublesome acid?' Such seemingly plausible ideas just won't work, because no incineration or neutralization process can have any effect at all on radioactive nuclei. The chemical form might be radically changed, but all the radioactive isotopes would continue to decay at their normal rates. To have any influence on radioactivity extreme conditions are required, for example bombardment by high-energy particles. Such measures might indeed produce changes in the radioactive nuclei, transmuting them into nuclei of other elements that are stable or have shorter half-lives. But it is doubtful that the process could be tailored to produce the right transformations of all the isotopes in a complex mixture, and in any event the expense would be prohibitive. Except possibly for special purposes, like deactivating tiny amounts of a single isotope, transmutation in high-energy accelerators holds no hope for ameliorating the problems of waste disposal.

1.4 KINDS OF RADIOACTIVE WASTE

Radioactive materials are widely used in industrial processes, in research laboratories, in hospitals and clinics, and in the preparation of fuel for nuclear reactors. All of these uses generate waste containing radionuclides, and the waste takes many forms. None of it can be simply discarded, like ordinary laboratory debris. Solutions can't be poured down the sink, paper towels and gloves can't be thrown in waste baskets, broken glassware can't be mixed with other rubbish. Any object, any material that has had contact with radioactive substances must be set aside for special handling.

Other kinds of waste come from the operation of nuclear reactors, and these also have many forms. Reactor wastes range from very dilute solutions to the intensely radioactive fuel elements from the reactor that are discarded after use. Even some of the gases produced by a reactor constitute a form of radioactive waste. The various kinds of waste pose different problems of disposal; techniques suitable for one kind may be useless or risky for another. All techniques of course, have a single goal, to ensure that the radioactive constituents of the waste are not released into human environments, or that the releases, if any occur, will be kept to very small amounts.

Current plans for most varieties of waste call for their disposal in geologic environments, where the natural properties of rock and soil are expected to play a major role in controlling the release of radioactive material. Disposal sites will be monitored for some time after the waste is emplaced, to make sure that isolation of the radionuclides is effective. Apart from this, it is hoped that the waste will need no further human attention. Once disposed of, the waste is expected to remain in place for a very long time, except for possible minor releases that will not increase radiation levels by more than the observed variations in levels of natural radiation.

In all geologic disposal schemes the most serious question about the effectiveness of long-term isolation relates to the possibility that water at some time in the future may reach the waste and dissolve radionuclides in unanticipated quantities. In the following chapters this possibility is critically examined for various kinds of waste and for various proposed or existing disposal methods. Emphasis however, is on the most dangerous kind of radioactive waste, the high-level material generated in nuclear reactors. This is the kind of waste that has the greatest potential for harming the environment, and that therefore is the focus of major public attention and major disposal efforts.

2

High-level waste:
the problem

2.1 GENERATION OF HIGH-LEVEL WASTE

In a rational world it would seem convenient to classify the many kinds of radioactive waste as high-level, medium-level, and low-level, according to the intensity of radiation they emit. This is sometimes done, but there is no agreement as to where the boundaries should be drawn. More commonly, but not very logically, the designation 'high-level' is restricted to waste produced directly by operation of a nuclear reactor, with or without subsequent chemical treatment, and all other waste is lumped into 'low-level'. It is the high-level waste (HLW) from reactors that poses the most troublesome problems of geologic disposal, and to these problems the next seven chapters are devoted. As a preface to the problems, we look first at the ways in which different kinds of HLW are generated.

Production of energy in a nuclear reactor depends on a curious property of one of the isotopes of uranium. Atoms of this isotope, ^{235}U, when struck by slow-moving neutrons, undergo *fission*, which means that their nuclei split apart into fragments with roughly half the atomic mass of the originals. (Note that this is very different from radioactive decay, in which an atom spontaneously loses a particle whose mass is only a tiny fraction of the original.) Fission generates enormous amounts of energy. It also liberates additional neutrons which, if going at the right speed, can then cause fission of other ^{235}U atoms. Thus a chain reaction is possible (Figure 2.1), each step of which produces more energy than the preceding because more atoms are involved. Cascading reactions of this sort are responsible for the huge release of energy in the explosion of a nuclear bomb.

^{235}U is of course a constituent of natural uranium. One might then reasonably inquire why uranium ores don't simply explode and disappear. A few stray neutrons are always present in ordinary materials; a few ^{235}U atoms in any specimen of ore must be undergoing fission; why doesn't the reaction propagate itself through the ore until all the ^{235}U is gone? Several factors prevent such a catastrophe. For one thing, ^{235}U is a minor isotope,

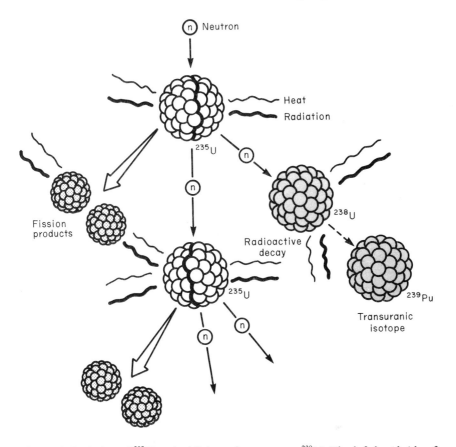

Figure 2.1 Fission of ^{235}U and addition of neutrons to ^{238}U. The left-hand side of the diagram shows the splitting of a ^{235}U nucleus by slow neutrons, resulting in release of energy and generation of more neutrons. On the right-hand side a fast neutron adds itself to a ^{238}U nucleus, producing the nucleus of a heavier element. (The diagram suggests direct production of plutonium; this is an over-simplification, because intermediate steps are involved.) (Office of Technology Assessment, U.S. Congress)

making up only about 0.7% of natural uranium. Secondly, most of the neutrons produced by fission cannot reach other ^{235}U atoms because they escape into air or are absorbed by other constituents of the ore (including the more abundant uranium isotope, ^{238}U, which does not undergo appreciable fission), and thirdly, most of the fission-generated neutrons are moving too fast to cause fission, even if they do strike other ^{235}U atoms.

For the chain reaction to persist, special conditions are needed. One way to set up such conditions is to recover uranium from its ore, increase its content of ^{235}U, and then embed chunks of the enriched uranium in a substance (moderator) that will slow neutrons without absorbing too many of them. Atoms of several light elements are effective moderators: hydrogen (in the form of water, H_2O, or deuterium oxide, D_2O, or some other simple compound) and carbon (in the form of graphite) are commonly used. When a certain minimum amount ('critical mass') of this mixture of uranium-plus-moderator is brought together (by rapidly moving subcritical masses into contact), the mixture will spontaneously explode. This is the recipe for one kind of nuclear weapon.

If fission is to produce useful energy rather than a mushroom cloud, the chain reaction must be controlled. This is accomplished by inserting between the uranium chunks a substance like boron that is an efficient absorber of neutrons. If the amount or position of the absorber is adjustable, the reaction can be kept going at whatever rate of energy production is desired. An electric power reactor, then, has three essential components: uranium, in the form of metal or a compound, to serve as fuel; water, or deuterium oxide or graphite, to moderate neutron speeds; and boron (or other material) to act as a control. (In some kinds of reactors water plays a dual role, serving both as a moderator to slow neutrons and as a control to absorb some of them.) The energy given out by the controlled fission is used to heat water, forming steam which operates a turbine, which in turn drives an electric generator.

Reactors have many designs, but in the most common kind the fuel has the form of small uranium dioxide pellets placed in metal tubes (usually made of zircaloy, an alloy of zirconium) 3 or 4 meters long and a centimeter or so in diameter. These long, slender fuel rods (or fuel pins) are mounted in clusters (Figure 2.2) and placed in a large tank of water. If a substance other than water is used for control, rods of this substance are set up so that they can be lowered into or between the fuel-rod clusters, as needed to keep the neutron flux at the proper level. With this arrangement the reaction starts spontaneously and maintains itself for a long time, the fission process in each fuel rod producing neutrons that travel through water and are slowed sufficiently to cause fission when they strike ^{235}U atoms in adjacent rods.

The products of fission – the atoms of elements with roughly half the atomic mass of uranium – accumulate in the fuel rods as the reaction goes on. These atoms have considerable variety, representing elements in the middle of the periodic table (selenium, cobalt, strontium, and cesium are examples), but the atoms belong to radioactive isotopes of these elements rather than the familiar stable forms. As the quantity of radioactive fission products increases, they interfere with the fission reaction by absorbing and deflecting neutrons. After a year or so of operation the interference is so serious that energy production is no longer efficient. To keep the reaction going, the fuel rods must then be removed and replaced by new ones. The

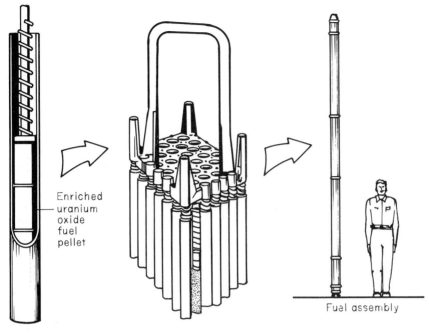

Enriched
uranium
oxide
fuel
pellet

Fuel assembly

Fuel rod

Figure 2.2 Fuel rods. The left-hand drawing is a cut-away of part of a single rod, the middle drawing shows the upper end of a cluster of rods, and the right-hand picture gives the scale of a complete fuel-rod cluster. Many such clusters in a tank of water constitute the energy-producing part of one kind of nuclear reactor. (U.S. Department of Energy)

old rods are intensely radioactive and thermally hot, so that all handling must be by remote control. To prevent them from contaminating the environment, they are lifted out of the reactor and placed in large basins of circulating water (Figure 2.3). (Dry storage – enclosing the fuel rods in metal casks and spacing the casks on a concrete foundation to ensure circulation of air – is an alternative under consideration in Canada and the United States and already used to some extent in West Germany.) The water cools the rods, and together with materials of the basin absorbs the radiation they emit. The rods are no longer of any immediate commercial value. They are highly dangerous if an accident should release them from the water basin, and they are a nuisance to the power company that must take care of them. They constitute one form of high-level waste.

A second form of HLW results from a different process that goes on in the fuel rods during reactor operation. The heavy isotope of uranium, ^{238}U, which makes up the great bulk of material in the fuel, does not fission like

Figure 2.3 Spent fuel-rod clusters in temporary storage in a water basin constructed of concrete and steel. Fuel rods in water basins at reactor sites are one form of high-level nuclear waste. (U.S. Department of Energy)

^{235}U, but some of its atoms add neutrons to their nuclei to form atoms of ^{239}U. These are unstable, and by radioactive decay and addition of further neutrons are transformed into atoms of other elements, all with atomic masses greater than that of uranium (Figure 2.1, right-hand side). These heavy elements are called collectively the *transuranic elements*; they include neptunium, plutonium, americium, curium, and minor amounts of several others, all of them elements whose existence was unsuspected before 1940. All isotopes of the transuranic elements are radioactive, many are alpha-ray emitters, some have very long half-lives. They accumulate in the fuel rods along with the fission products and contribute to their radioactivity.

One of the heavy elements, plutonium, has an isotope of special interest.

This is ^{239}Pu, whose atoms have the same property as ^{235}U: they split apart when struck by neutrons, give out energy, and liberate additional neutrons. Thus ^{239}Pu, like ^{235}U, can be used in bombs, and has proved to be even more suitable for this purpose. The element is therefore of great interest to military establishments, and in a few countries large amounts have been produced. To obtain plutonium the spent fuel rods from a reactor, instead of being stored in a water basin, are broken into small pieces and placed in a bath of nitric acid. The acid dissolves all the radioactive material – the remaining UO_2, the fission products, and the heavy elements formed from ^{238}U. From this solution the plutonium is separated by addition of an organic chemical. The remaining nitric acid, containing all the other radioactive elements, has no further use and must somehow be kept from contaminating the environment. The common practice is to pump it (by remote control, of course) into huge double-walled steel tanks, sunk a few meters beneath the ground surface (Figure 2.4). This hot acid solution is the second major form of high-level waste.

Because its production involves the re-handling of spent fuel rods, such material is called reprocessing waste. Its acidity is commonly neutralized by addition of alkali, which causes some of the solutes to precipitate, and evaporation leads to partial crystallization of other constituents. Thus the tanks of reprocessing waste contain complex mixtures of liquid and solids, all of them radioactive. Large tank farms of this material, spoken of euphemistically as 'defense waste', have grown near the Hanford plant in

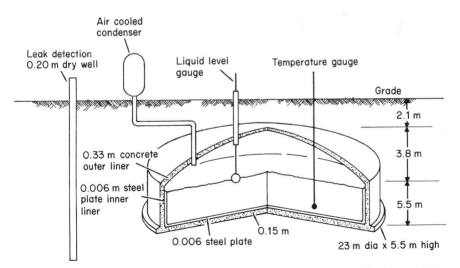

Figure 2.4 Diagram of a tank for holding reprocessing waste. (Based on U.S. Department of Energy)

south-central Washington and the Savannah River plant in South Carolina, the two places where plutonium for military purposes has been produced in quantity in the United States (Figure 2.5).

The fissile isotope of plutonium could be used in place of ^{235}U in power reactors, as well as in nuclear weapons. Some of it is so used, perforce, in existing reactors: as ^{239}Pu forms by neutron reaction with ^{238}U during reactor operation, it is subject to the same bombardment by slow neutrons as is ^{235}U, and accordingly a part of it undergoes fission. This reaction, in fact, is responsible for about a third of the power generated in an ordinary reactor. The energy obtainable from a given quantity of uranium fuel however, would be markedly increased if the unfissioned ^{239}Pu remaining in the fuel rods after their removal from the reactor could be recovered. This would mean dissolving the fuel rods, just as is done for military purposes, and chemically separating the plutonium. In the same operation any remaining ^{235}U could be separated also, and with the recovered ^{239}Pu could then be fashioned into new fuel rods. This procedure would eliminate the need for prolonged storage of spent fuel in water basins. There would then be only a single kind of high-level reactor waste: a

Figure 2.5 Tanks for reprocessing waste under construction at Hanford, Washington. When completed, the tanks will be covered with a meter or two of earth. The liquid-plus-solid mixture in tanks of this sort is the second kind of high-level waste. (U.S. Department of Energy)

neutralized nitric acid solution like that now present in the tanks at Hanford and Savannah River.

For maximizing energy production, the reprocessing of spent fuel to recover the fissile isotopes of plutonium and uranium makes good sense. In the early days of reactor development it was assumed that this procedure would be generally followed. But the idea ran afoul of political concerns: if reprocessing of spent fuel became common practice, many plants producing plutonium would be set up, and the plutonium would be difficult to guard from pilfering by would-be terrorists bent on using it for unsavory purposes. For this reason (and also economic reasons), reprocessing of commercial spent fuel has not become widespread. Reprocessing on a modest scale has been started recently in England, France, West Germany, and Japan; some other countries may elect to ship their spent fuel to these four. In the United States, for both political and economic reasons, no commercial reprocessing is planned for the near future; Canada has deferred a decision about reprocessing until the next century. The political argument in opposition is still potent, although many do not find it compelling. For the more distant future it can be plausibly urged that ultimately the world will need the very considerable amounts of energy remaining in the spent fuel rods that are accumulating at reactor sites, and that therefore the reprocessing option, however dubious it may seem at present, should remain open.

The controversy over reprocessing means that spent fuel at present has an ambiguous status (Figure 2.6). From one point of view it is radioactive waste, but from another it is a valuable energy resource for the future. Because the first viewpoint is the commoner one and because plans are far advanced for the treatment of spent fuel as a form of waste, it will be considered waste for purposes of this book. But ideas about waste change rapidly, and the possible use of spent fuel for energy production may well seem more important at a later time.

In general then, two principal kinds of HLW are accumulating: spent fuel rods in water basins at sites of electric-power reactors, and reprocessing waste in steel tanks at places in a few countries where plutonium has been or is being produced. Both kinds of waste, it should be emphasized, are at present adequately isolated from the environment. Radiation is absorbed by materials of the water basins and the tanks and by the surrounding soil, and temperatures are controlled when necessary by artificial cooling. Long-term monitoring of the sites where the waste is stored has shown no appreciable increment to natural radiation levels, except on rare occasions when minor leaks have occurred. Thus one may fairly ask, 'Why then, is waste disposal such a troublesome matter? Why not just leave the stuff where it is, and not worry about disposal at all?' This is a reasonable question, but it does not take into account the problem of time.

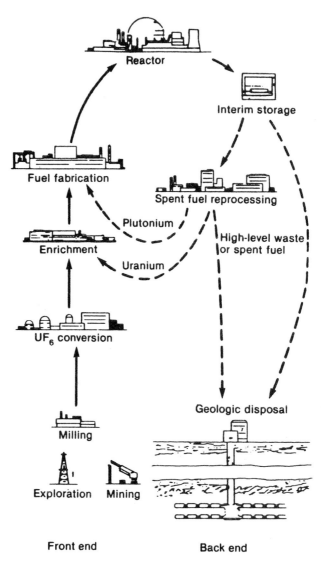

Front end

Back end

Fuel cycle today
Prospective "closed" fuel cycle

Figure 2.6 Diagram of the commercial nuclear fuel cycle, showing the origin of the two kinds of high-level waste destined for ultimate geologic disposal. Fuel discarded from the nuclear reactor goes to 'interim storage' in water basins, and from there either directly to the disposal facility or to a plant where plutonium and uranium are separated and reprocessing waste is generated. (U.S. Council on Environmental Quality)

2.2 STORAGE, DISPOSAL, CONTAINMENT, AND ISOLATION

Time enters the picture because the waste, in its present situation, must be subject to constant surveillance. As long as all equipment remains in good shape and care is taken to detect leaks promptly and remedy them when they occur, existing waste can indeed be kept safely in its present condition for an indefinite time. But equipment does not last for ever: the metal of containing tanks corrodes, cracks unexpectedly develop, pumps for cooling water wear out. When such mishaps occur, someone must be on hand to build new tanks and install new pumps. For how long must such surveillance be maintained?

A precise answer is elusive, but the time is certainly long in comparison with human life. Radioactive waste remains potentially hazardous for varying times, depending on the kind of radionuclides it contains. The cobalt isotope ^{60}Co, common in medical waste, decays to harmless levels within a few decades. ^{90}Sr and ^{137}Cs, which are abundant in much reactor waste, require at least a few centuries. Some isotopes of the transuranic elements, plus ^{129}I and ^{99}Tc, remain potentially hazardous for more than ten million years – if all or an appreciable quantity should be released from storage at once. It is hard to set a specific moment when surveillance over stored HLW could be relaxed, but the time is obviously so far in the future that maintenance of oversight cannot be expected. Continued surveillance requires continuity of social and political institutions; in the 6000 years of recorded human history, how many institutions have a record of lasting more than a few centuries? Clearly we cannot count on our descendants to maintain for a million years and more an interest in watching over the waste that we have generated.

This then, is the argument for finding a more permanent method of disposal – and a very convincing argument it is. The waste is in safe storage now, but it will not remain safe unless someone takes care of it, and we cannot be sure that adequate care will continue. We need a place to put the waste where it will be far from surface environments and will require no further watching. For this purpose, it is generally agreed, the waste should be moved deep underground. To ensure stability, it should all be converted to a solid form before burial; the fuel rods, of course, are already solid, but reprocessing waste must be further treated to change it to a solid, presumably some kind of glass. The problem of disposal is to find a site for the deep burial of these solids, and a method of emplacement that will ensure adequate isolation of the waste, with no need of further human attention, for a long time into the future.

What meaning can we give to 'adequate isolation'? Ideally the waste should be so well entombed that none of it will ever reappear at the earth's surface. But this is probably a vain hope: materials are not available that will enclose the waste and endure for the necessarily long periods of time, and we cannot count on finding geologic environments where access of ground water to the waste is excluded for ever. We should be able however, to contain the waste completely for the shorter times (say a few centuries) needed for decay of the more active nuclides (especially ^{90}Sr and ^{127}Cs), and to ensure that thereafter the dissolution and movement of radioactive material in ground water are sufficiently impeded that only minute amounts will find their way to the surface. This is the strategy of current plans for disposal: to prevent any escape of radionuclides for a time, then to keep the waste so effectively isolated that no more than the traces permitted by regulatory agencies can reach the environment, even during the times needed for decay of the longest-lived isotopes.

In this discussion of time it should be noted that a few words have been used which are peculiar to the jargon of the waste-disposal enterprise, familiar words that take on meanings not quite the same as in ordinary usage. *Storage*, for example, is distinguished from *disposal*, in that the former refers to temporary retention of waste, with the expectation that it will later be moved somewhere else, while the latter connotes emplacement with no intention of later recovery; thus spent fuel is now *stored* in water basins, with the expectation that sometime it will be *disposed of* permanently in repositories underground. An objective of disposal may be either the *containment* of waste, which means the retention of all toxic material within a designated boundary, or the *isolation* of waste, meaning retention to the degree that any escaping radioactive material will be kept within prescribed regulatory limits. Thus in current plans the waste in a repository will be *contained* for at least a few centuries after the repository is filled, and thereafter will be *isolated* in the sense that only minute amounts can escape.

2.3 IS IMMEDIATE DISPOSAL NECESSARY?

About the need for ultimate disposal of nuclear waste there is little dispute, but the urgency of the need can excite a lively debate. The debate involves both technology and politics, and is often confusing to the public.

On the one side are those who see HLW as a major menace, and chafe at any delay in getting it underground. As long as waste is stored in tanks near the surface, there is always a chance of catastrophic leaks – from corrosion, earthquakes, a plane crash, or terrorist activity. The necessary technology for disposal is at hand, except perhaps for details. Why not get started at once on the long job of building an underground repository for this dangerous material? Besides, there are lingering doubts that geologic disposal is feasible; shouldn't an immediate effort be made to demonstrate that it can indeed be accomplished? Furthermore, protracted delay will mean that the problem will be passed on to future generations; isn't it more equitable that we in this generation take care of our own nuclear waste now, rather than leave it for our grandchildren to cope with? The argument for immediate action is a strong one, and in the United States it has led to a sense of great urgency.

On the other side are voices that counsel slowness, for reasons at least equally good. For one thing, the steady decay of radioactive atoms means that HLW becomes less radioactive and less thermally hot as time goes on; after a few decades it is easier to handle, less hazardous to operating personnel, less potentially destructive to the rock environment in which it will be placed. Moreover, one can confidently expect technological advances in the near future that will make disposal easier and safer. For spent fuel there is the additional argument that recovery of its contained plutonium and uranium may ultimately seem desirable to augment energy supplies, so that this kind of HLW should be kept accessible near the earth's surface rather than buried deep below. In many European countries these arguments have prevailed, and actual disposal operations have been postponed for at least several decades. Even in the United States, despite widespread clamor for getting disposal under way, the first waste to be buried will perforce be many decades old, simply because of the manifold delays that have beset and are besetting the waste disposal program.

A confirmed skeptic can even build a reasonable case for doubting that permanent disposal is a wise course, however long it may be delayed. At present HLW is simply a nuisance that we want to get rid of, but can we be sure that future generations will regard it in this light? Possible recovery of fissile plutonium and uranium is one obvious reason for which spent fuel rods might be of interest to people in the future, but HLW might become attractive in other ways as well. The waste is giving out large amounts of heat energy; with present technology there is no way to put this energy to work (despite much effort to find uses for at least part of it), but can we be sure that technological progress will not someday make the energy usable? Also, HLW is most interesting material metallurgically, since it contains large amounts of isotopes of rare metals like cesium, zirconium, and technetium, as well as the transuranic elements; their radioactivity precludes their separation and use at present, but is it certain that this will always be true? It may well be, according to our skeptic, that with improvements in technology HLW will become a valuable mineral resource for our descendants. Why then, go to the trouble of putting it deep underground? Why not store it, with proper safeguards, in a structure at or near the earth's surface, where its behavior can be watched and where it will be accessible to people of the future who may find uses for it?

The argument is hard to refute, except on political grounds and on the philosophical premise that we who benefit from nuclear energy today should not leave the handling of its unpleasant debris as a problem for our remote progeny. The political drive to demonstrate that disposal of HLW can in fact be accomplished is so strong, at least in the United States, that our skeptic's cautionary note is lost in the chorus of voices debating the merits of various geologic disposal sites and various ways to assure the necessary isolation. If in the chapters to come this clamor of voices is the focus of attention, the possible merit of an opposing viewpoint – that the whole enterprise may be a horrendously expensive mistake – should not be completely forgotten.

2.4 AMOUNTS AND COMPOSITION OF HIGH-LEVEL WASTE

From an engineering standpoint, how difficult will the problem of disposal be? How much waste is there to dispose of? How fast is the quantity increasing and what problems will arise because of its peculiar composition?

The quantity of HLW already existing in the United States, and world-wide estimates of amounts to be produced by the year 2000, are shown in Table 2.1. Missing from the table is reprocessing waste in countries other than the United States; the amount is small except in the Soviet Union, and there it is probably comparable to the United States figure. Actual quantities of material to be placed underground will be greater than the numbers in the table, because the waste will be enclosed in heavy metal canisters and other substances will be added to at least some parts of it. Details of the planned treatment of HLW for burial are still uncertain. Spent fuel rods may be left intact, which would require canisters 4 or 5 meters long, or they may be broken up and embedded in a metal or ceramic matrix. Reprocessing waste will be solidified, probably by adding materials and heating so as to incorporate the waste in glass; various compositions have been suggested for the glass, but currently the one favored in most countries is a borosilicate glass that experiment has shown to be resistant to dissolution in simulated repository environments. Some reprocessing waste from power reactors is being vitrified in this manner in Europe and Japan, and large-scale vitrification of the waste from plutonium production is expected soon in the United States. Additions of material to the waste in such activities will presumably at least double the total amount of material to be handled.

Even so, the quantity is small compared with amounts of other kinds of chemical waste produced by modern industry. For example, the amount of spent fuel generated annually in the United States is between 1000 and 2000 tonnes, compared with an estimated 260 000 tonnes of other hazardous waste. To visualize the magnitude of the disposal problem, imagine that the

Table 2.1 Amounts of high-level waste

United States				
	Total accumu-lated and in stor-age in 1984	Radioactivity	Total estimated to be in storage in 2000	Radioactivity
Spent fuel	11 700 tonnes (5 300 m³)	50.3×10^{13} MBq†	41 600 tonnes (18 700 m³)	135.0×10^{13} MBq†
Reprocessing waste	370 000 m³	4.8×10^{13} MBq	327 000 m³	5.6×10^{13} MBq

Other countries
Total amounts of spent fuel (tonnes) estimated to be produced by the year 2000, for all countries where the amounts are greater than 1000 tonnes:

Argentina	5 800	East Germany	2 100	South Africa	1 200
Belgium	3 000	*West Germany	11 000	*Spain	5 100
Bulgaria	2 500	Hungary	1 400	Sweden	5 000
Canada	38 000	India	5 000	Switzerland	2 000
China	1 300	Italy	3 700	Taiwan	2 600
Czechoslovakia	3 800	*Japan	21 000	*United Kingdom	38 000
Finland	1 400	Korea (South)	4 400	*USSR	12 600
*France	37 000	Romania	8 200		

*In these countries the amounts of spent fuel in storage will be less than the numbers indicate, because part of the fuel is being reprocessed and the waste converted to glass.
†1MBq = 1 megabecquerel = 19^{9}Bq = 2.7×10^{-5} curie
Sources: For United States, U.S. Department of Energy, *Spent fuel and radioactive waste inventories, projections, and characteristics*, DOE/RW-0006, Rev. 1, December 1985. For other countries: Harmon, K.M. and Johnson, A.B., *Foreign programs for the storage of spent nuclear power plant fuels, high-level waste canisters and transuranic wastes*, prepared for the U.S. Department of Energy by Pacific Northwest Laboratory of Battelle Memorial Institute (PNL-5089, UC-85), 1984.

existing waste in the United States alone, both fuel rods and reprocessing waste (Table 2.1), could be dumped into a college football stadium: the pile would fill the stadium to a depth of 8 to 10 meters. To get this volume of radioactive junk put into canisters and moved underground, with all the handling by remote control, is certainly a major engineering project. But it is by no means unreasonable, and it is dwarfed by the ever-increasing problem of safely managing the enormously greater volume of ordinary industrial waste.

The composition of HLW has a wide range, depending on the characteristics of the reactor from which it comes, on the kind of fuel used, on the length of time the reactor operates, and on the time since the waste was removed from the reactor. Reprocessing waste from plutonium production differs obviously from spent fuel in its lower quantity of plutonium isotopes; if uranium is also recovered during reprocessing, to be used in new fuel rods, this element also will be greatly reduced in the waste. Any of the waste contains a great variety of radionuclides, some from the fission of ^{235}U and some from the addition of neutrons to ^{238}U. Some idea of the quantities can be gained from Table 2.2, but it should be noted that these analyses are for specific samples of waste and might be substantially different for other samples. In addition to the radioactive isotopes, the different kinds of waste contain large amounts of inert material – notably zirconium in the case of spent fuel (from the tubes used to contain the fuel pellets), and a variety of substances from the reprocessing and glass-making operations in the case of reprocessing waste.

In planning for disposal, the major radioisotopes of concern are: the two fission products ^{90}Sr and ^{137}Cs, which decay to innocuous levels within a few centuries; the long-lived fission products ^{129}I and ^{99}Tc; and several

Table 2.2 Important radionuclides in two kinds of high-level waste

Isotope	Half-life	In one spent-fuel assembly from a pressurized water reactor		In one canister of borosilicate glass made from reprocessing waste	
		Time after discharge from reactor		Time after reprocessing	
		10 years	1000 years	10 years	1000 years
^{90}Sr	28.8 yr	200g	0g	880g	~0g
^{99}Tc	2.14×10^5	360	360	1600	1600
^{129}I	1.6×10^7	83	83	0.38	0.38
^{135}Cs	3×10^6	140	140	2400	2400
^{137}Cs	30.2	440	~0	2000	~0
^{234}U	2.45×10^5	88	150	2.9	16
^{238}U	4.47×10^9	440 000	440 000	9900	9900
^{237}Np	2.14×10^6	210	660	930	1000
^{238}Pu	87.7	60	260	13	0.014
^{239}Pu	2.4×10^4	2300	2300	53	68
^{240}Pu	6570	1100	1000	36	56
^{242}Pu	3.76×10^5	210	210	4.8	5.0
^{241}Am	433	230	120	120	26
^{243}Am	7370	40	36	180	160

Source: NRC/NAS (1983) *A Study of the Isolation System for Geologic Disposal of Radioactive Waste*. Washington: National Academy Press. pp 31 and 35.

long-lived isotopes of the actinide elements U, Np, Pu, Am, and Cm. Two additional isotopes that are very minor in fresh waste but would become important after several thousands of years are ^{226}Ra and ^{210}Pb, generated by decay of actinide elements in the waste. The proportions of all these radioisotopes change with time, because of their different rates of decay. An idea of the potential toxicity of the various isotopes, and the way the relative toxicities change over time, can be gained from Figures 2.7 and 2.8, which show the 'water-dilution volumes' for radioisotopes in the amount of spent fuel discharged yearly from a 1-GW(e) pressurized-water reactor and in the waste that would be generated from this spent fuel if it were reprocessed. The water-dilution volume for each isotope is the volume of water that would be needed to dilute the amount of the isotope in the waste to a concentration safe for ordinary use. The greater the volume of water

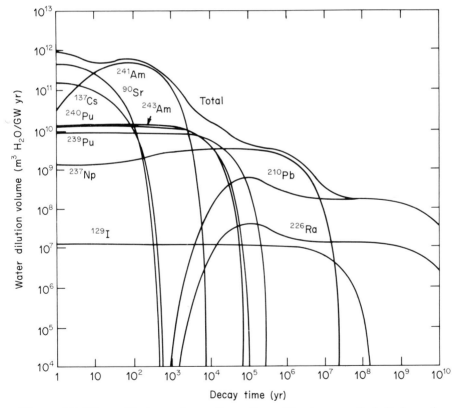

Figure 2.7 Water-dilution volumes for radionuclides in spent fuel discharged from a 1-GW(e) pressurized-water reactor. (Choi, J.-S. and Pigford, T.H. (1981) Water-dilution volumes for high-level wastes, *Trans. Am. Nucl. Soc.*, **39**, 176–7)

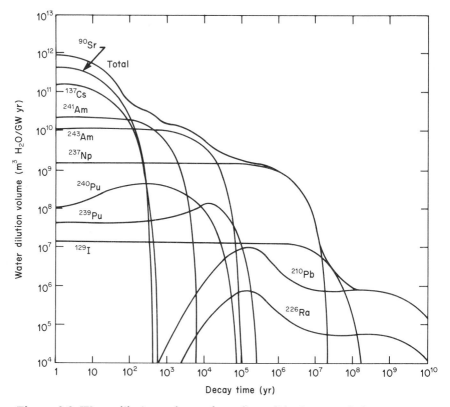

Figure 2.8 Water-dilution volumes for radionuclides in spent-fuel reprocessing wastes formed by operating a 1-GW(e) pressurized-water reactor for one year. (Source as for Figure 2.7)

needed, the more dangerous the isotope would be if all of it were suddenly released. The water-dilution volume for a given isotope decreases with time, at a rate depending on its half-life (although for a few the volume shows a temporary increase, because these isotopes are in part generated in the waste by decay of other radionuclides). Thus different isotopes become the principal contributors to toxicity of the waste at different times – first ^{90}Sr and ^{137}Cs, then for most of the first 20 000 years two isotopes of americium, then ^{237}Np, and after about 10 million years ^{210}Pb and ^{226}Ra. The general sequence of toxicities is much the same for the two kinds of waste, but plutonium isotopes are of course more important contributors to toxicity in the unreprocessed spent fuel.

The problem of disposal then, is how to place safely underground the quantities of waste indicated by Table 2.1, given that the waste contains an assortment of radionuclides like those listed in Table 2.2, and that the amounts of the nuclides are large enough to make the waste toxic, if it should escape, for a very long time into the future. Can we predict for such very long periods of time just how much of the waste is likely to escape from a geologic repository? This is a query to daunt the most optimistic of geologists, for it is a new kind of question: geologists are accustomed to making accurate reconstructions of the past, but not to forecasting the future in any detail. The current status of attempts to answer the question is the theme of succeeding chapters.

__3__

Strategies for solving the problem

3.1 ALTERNATIVES FOR DISPOSAL

We are assuming that the method for taking care of HLW will be mined geologic disposal – burial of the waste in a cavity excavated in bedrock deep underground. This is the method currently being planned in all countries where HLW is accumulating. The decision to use this means of disposal however, was reached only after long consideration of other possible techniques. Some of the alternatives seem attractive, but they have all been abandoned – or at least deferred for possible later study – either because they were prohibitively expensive, because they entailed apparently greater risk, or because the present state of technical knowledge is not sufficient to evaluate fully their consequences. Geologic disposal has the advantages that it can probably be accomplished at a tolerable cost and that the necessary knowledge to predict its results is probably adequate. A few gaps in the information base remain, but seemingly they can be filled by research that is under way or planned.

A brief look at some of the alternative suggestions for HLW disposal is useful by way of illustrating the present state of technical knowledge and of highlighting the merits of the planned geologic method. We discuss five of the more promising alternatives, then contrast them with the salient features of mined geologic disposal.

Disposal at or near the surface

First is the suggestion mentioned in Chapter 2 that HLW with proper shielding could be kept near the earth's surface rather than moved underground, thus making it easier to monitor and more accessible if its recovery should seem desirable at a later date. Shielding much more substantial than that provided by the present tanks and water basins would be needed. Two ways of providing adequate shielding have been proposed: entomb the waste in a monolithic structure (like the Egyptian pyramids for example, but not quite so ponderous), or put the waste at the end of a horizontal tunnel driven into a mountainside. The obvious worry about any position near the surface is the greater vulnerability of the waste, over a very long time, to human or natural disturbance – weathering and erosion, earthquakes, a plane crash, or intrusion by political dissidents seeking fissionable material for ulterior ends. One can argue that this concern is exaggerated, but it has been sufficient to discourage further consideration of surface disposal.

Disposal in space

Putting HLW in rockets and shooting the rockets into outer space or into orbit around the sun is an alternative with strong appeal, because obviously complete elimination from terrestrial environments would be guaranteed. The expense would be astronomical if all existing and future waste were handled in this manner, but conceivably small volumes of the more toxic and long-lived constituents of HLW could be separated out to be placed in rockets, while the bulk of the waste would remain as lower-level material for which terrestrial disposal should be a simpler problem. One objection comes to mind at once: rocket-launching is not foolproof, and explosion of a rocket on lift-off might scatter radioactive debris over a wide area. Advocates say in response that if expense is no object, the capsule containing waste can probably be made sturdy enough to survive a rocket mishap. A more cogent objection lies in doubts that the procedure would be worth the trouble: separating waste into two fractions would be a diffi-cult and expensive operation, hazardous to the personnel involved; and the large volume of remaining lower-level waste, although easier to handle than the original, would still be a major disposal problem. Perhaps rocket disposal will be used at some time for special kinds of concentrated waste, but as a general method for dealing with HLW it does not seem practicable.

Disposal in an ice sheet

The thick ice sheets that cover Greenland and Antarctica look like fine places to entomb HLW, far from any possible contact with inhabited parts of the globe. Just take the waste to the middle of the ice, set it down on the surface, and let it melt its way downward through the ice sheet to the bottom. At first glance it would seem to be safe there for all eternity, or at least a considerable fraction thereof. But then one starts to wonder: what if heat from the waste melts the bottom layers of ice, lubricating flow of the ice sheet towards its margins? How long would it take the moving ice to carry some of the waste to the surrounding ocean? Even if the waste doesn't move bodily, wouldn't some of the radionuclides be dissolved and carried under the ice in solution? Answers to these questions we simply do not have; the long-term effects of a potent heat source at the base of an ice sheet are hard to predict. Added to this uncertainty is the logistic difficulty of getting large volumes of waste to the middle of an ice sheet, plus the political problem of convincing nearby nations that the ocean lapping their shores will not be contaminated. The objections are so formidable that ice-sheet disposal does not seem attractive.

Disposal in a very deep hole

Rather than put waste only a few hundred meters under the surface, as is planned for ordinary geologic disposal, why not drill a very deep hole to a level where the rock is warm (3–5 km), put HLW at its bottom in the form of a slurry, and let the combined geothermal heat and heat from the waste melt the rock? The waste will then be incorporated in normal igneous magma, its radioactivity will be greatly diluted, and by the time the magma ultimately cools to solid granite and is exposed to erosion very little activity will be left. This is a particularly attractive idea, since it seemingly works with natural processes rather than in opposition to them. But again we face the inadequacy of present knowledge: can we be sure that the newly-formed radioactive magma will stay far below, where it is generated, and not rise to the surface to fuel a volcanic eruption? Even if the magma itself doesn't rise, might it not contaminate deep ground water or generate radioactive hydrothermal solutions? Or could it lead to earthquakes that would crack the rock above, providing channels for movement of radioactive magma and accompanying fluids? There are simply too many unknowns for this proposal to be considered seriously, at least for the present. Even if the unknowns were eliminated, the drilling of many holes to the necessary depth would be a very costly operation.

Disposal in the ocean floor

There is general agreement that nuclear waste cannot be safely disposed of by simply dumping it into the sea, except for some kinds of very low-level material. Burial within sediments at the bottom of the ocean however, is a possibility well worth considering. Two variants of this alternative have been suggested: put the waste in a deep ocean trough, where subduction can be expected to carry it slowly down into the earth's mantle; or bury it in sediment out towards the middle of one of the large oceanic plates, far enough from land so that slow motion of the plate cannot bring it near coastal waters before all the radionuclides have decayed to harmless levels. To the first variant an objection can be raised on the grounds that we don't know enough about the details of subduction to be sure that the sediments of any given trench will not be faulted up onto an adjacent continent, and hence ultimately exposed near the surface, rather than dragged down into the mantle. The second variant is more promising: surely the central part of an oceanic plate is as stable a geologic environment as could be found anywhere on the planet, and it is hard to see how waste canisters placed a few tens of meters under the ocean floor could be disturbed for millions of years. Still, a skeptic can point out that our knowledge of conditions at the bottom of the deep sea is far from complete. Would the waste containers corrode rapidly in contact with the interstitial sea water of the sediment? How much would the sediment be altered by heat and radiation from the waste? If waste is ultimately exposed to the salt water, how fast would radionuclides dissolve and migrate into the ocean above? If they do enter the ocean, would currents distribute them fast enough to pose a hazard to distant shores? These are worrisome questions, but they should be answerable by the kind of oceanographic research that is now in progress. Of all the alternative disposal methods, the subsea-bed option is the most promising, and the one to which the most research effort has been devoted. We will leave it for the present with this brief mention, but return to it for fuller discussion after the pros and cons of geologic disposal on land have been explored.

Mined geologic disposal

In contrast to these imaginative suggestions, geologic disposal seems disarmingly simple – and this is one of its great attractions. Just dig a hole, emplace the waste, and cover it up. This is the method that comes to mind, almost instinctively, for getting rid of any kind of disagreeable rubbish. From years of experience we know that natural processes will ultimately take care of many sorts of unpleasant substances when they are buried. Because radioactive waste is particularly dangerous, we would want to make sure it is buried at considerable depth, and we would be choosy about picking the site for its disposal so that geologic disturbances will not at some time uncover it. With these elementary precautions, the method would seemingly guarantee that the waste will never reappear to contaminate surface environments.

   More specifically, the concept of geologic disposal involves sinking a shaft to a depth of at least a few hundred meters, excavating an array of tunnels from the bottom of the shaft (Figures 3.1 and 3.2), encapsulating the waste in metal canisters (Figures 3.3 and 3.4), emplacing the canisters in holes drilled into the floors or walls of the tunnels (Figures 3.5 and 3.6), and backfilling tunnels and shaft with crushed rock or some sorbent material (Figure 3.7). All of these operations are either well-known mining procedures or operations that require only slight modification of standard

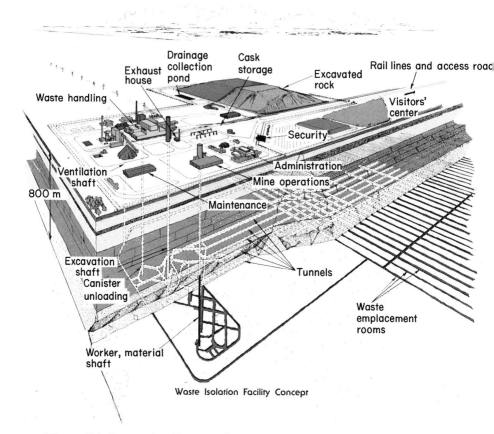

Waste Isolation Facility Concept

Figure 3.1 Perspective diagram of a mined geologic repository, showing surface installations, shafts, and array of tunnels. (U.S. Department of Energy)

engineering techniques. There are no apparent logistical difficulties in getting the waste to the disposal site, and no insuperable problems of emplacing the waste canisters, backfilling the tunnels, and sealing repository openings. Whereas other suggested disposal methods would require development of new techniques for transporting and emplacing waste, as well as research to answer questions about the probable behavior of the waste and its surroundings, geologic disposal seems to offer straightforward procedures and a minimum of unanswered questions. For this reason it is at present the method of choice.

Of course some questions remain, even for this most favored method. The plan calls for putting deep underground a large amount of material that is hot, highly radioactive, and far from chemical equilibrium with its surroundings. This has never been done before. Nothing in previous

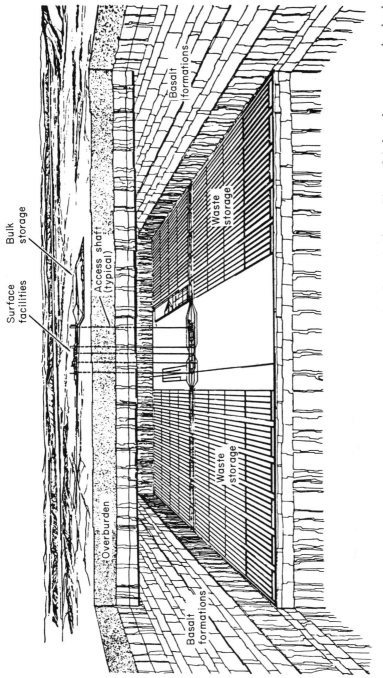

Figure 3.2 Cut-away perspective of a conceptual repository in basalt. This is similar to Figure 3.1, but shows more clearly the relation of the repository to the rocks in which it is excavated. (U.S. National Research Council (1983) *A Study of the Isolation System of Geologic Disposal of Radioactive Waste*, National Academy Press, Washington, DC)

Tunnel

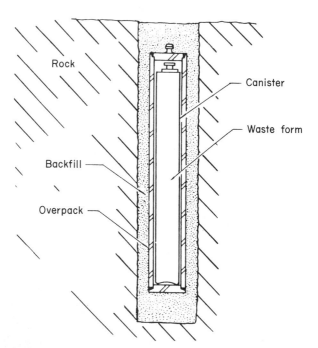

Figure 3.3 A waste canister emplaced in a hole drilled in the floor of a repository tunnel. The 'waste form' consists of either spent fuel rods or glass made from reprocessing waste; the 'canister' and 'overpack' are metal containers; the 'backfill' is crushed rock or clay or a mixture of the two. (Source as for Figure 3.2)

engineering experience can tell us just what to expect after the waste has been in place for a hundred, a thousand, or ten thousand years. How will prolonged exposure to heat and radiation affect the rocks enclosing the repository? How soon will the repository fill with ground water? How fast will the metal of overpack and canisters corrode? When ground water ultimately comes in contact with the waste, how fast will the various radionuclides dissolve? Will the dissolved substances travel freely through the rock with the moving ground water, or will they be delayed by precipitation and sorption on mineral surfaces?

Even though actual experience cannot give us answers to such questions, we do know enough about rocks and ground water to make plausible estimates, and ongoing research is adding steadily to our fund of knowledge. In contrast to the many questions that can be raised about alternative disposal methods, the uncertainties about geologic disposal

Figure 3.4 A typical canister for high-level waste. (U.S. Department of Energy)

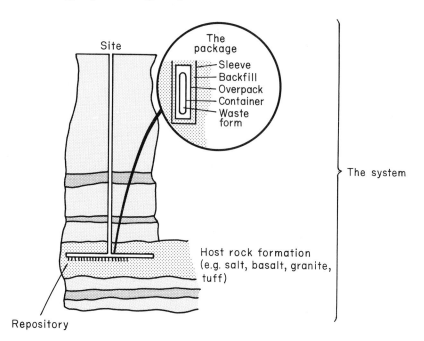

Figure 3.5 Component parts of a mined geologic disposal system, consisting of engineered barriers ('the package') and the principal geologic barrier ('host rock formation'). (U.S. Department of Energy)

seem readily approachable. Geologic disposal after all, involves only the uppermost part of the crust, a part of the planet with which we have long been familiar and about which reasonable predictions can be made. The kind of research needed is obvious, and only time and money remain as obstacles to carrying it out. The basic information here is well in hand, and further effort is needed only on details.

3.2 REQUIREMENTS FOR MINED GEOLOGIC DISPOSAL

If HLW could be sealed in canisters of an inert metal like gold or platinum, containment would be assured and the problem of disposal would be solved. Geology would play only a minor role, because such canisters placed in almost any kind of rock environment would last far beyond the time needed for all radionuclides to decay to harmless levels – provided only that a disposal site is chosen where disturbance by volcanic activity, tectonic movement, or deep erosion is unlikely. In the practical world a less durable metal must be chosen, and we must reckon with the almost certain ultimate corrosion of the canisters, contact of ground water with the waste,

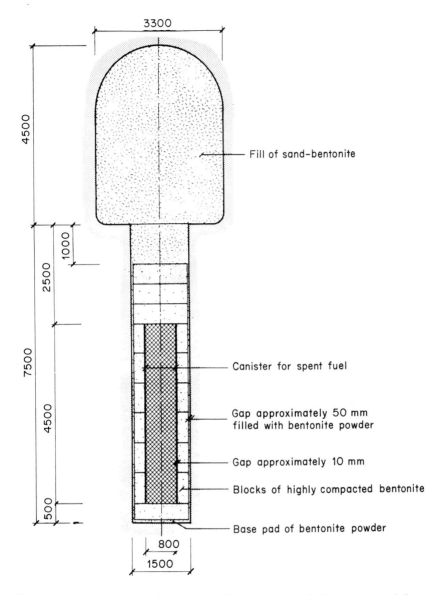

3300

4500

7500

2500

1000

4500

500

Fill of sand-bentonite

Canister for spent fuel

Gap approximately 50 mm
filled with bentonite powder

Gap approximately 10 mm

Blocks of highly compacted bentonite

Base pad of bentonite powder

800

1500

Figure 3.6 Cross-section of a tunnel and emplacement hole as proposed for a repository in Sweden. The canister is to be made of copper; the backfill pure bentonite clay in the emplacement hole and a bentonite-sand mixture in the tunnel. Dimensions are in millimeters. (Svensk Kärnbränslehantering AB, Stockholm)

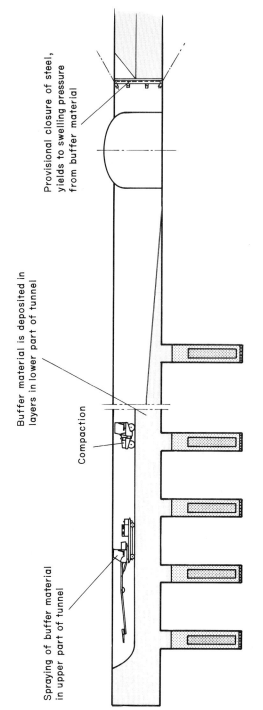

Provisional closure of steel, yields to swelling pressure from buffer material

Buffer material is deposited in layers in lower part of tunnel

Compaction

Spraying of buffer material in upper part of tunnel

Figure 3.7 Longitudinal section of a Swedish repository tunnel, showing the tunnel being backfilled with a sand–clay mixture ('buffer material') above the emplacement holes containing the waste canisters. (Svensk Kärnbränslehantering AB, Stockholm)

and dissolution of some part of its radionuclide content. This is acceptable provided that the onset of dissolution is delayed long enough for the more active isotopes to have largely decayed, and provided that thereafter the amount of radioactivity reaching the surface is kept very small.

The delay of dissolution is largely an engineering problem: the respository must be sealed and filled with material that will impede access of ground water, the canisters must be made of a metal resistant to corrosion (but more readily available than the precious metals), and the waste must be in the form of a highly insoluble solid. For the radionuclides that do eventually dissolve, preventing them from reaching the biosphere in more than minute amounts is a geologic problem: the repository site must be located where the quantity of ground water is small, its movement is slow, its composition is non-corrosive, and its path to the biosphere is long, and where the rock is capable of precipitating or sorbing most of the radionuclides that move through it. In short, the strategy of mined geologic disposal is to set up a series of barriers to the movement of radionuclides, barriers that are partly engineered and partly dependent on the geology of the disposal site.

How much reliance should be placed on engineering, and how much on geology? The question is largely rhetorical: some would put more emphasis on one kind of barrier, some on the other, but there is general agreement that both are needed. Because predictions about the effectiveness of any one barrier over very long times are necessarily uncertain, a number of barriers is desirable, each one made as effective as possible. Some will be redundant, but redundancy is preferable to uncertainty.

In qualitative terms the strategy seems clear enough, but obviously the requirements for a repository must be made more quantitative. The barriers set up to control movement of radionuclides must delay their escape for very long times; but just how long? Eventually some radioactive material will escape from a repository, and the amount must be kept very small; but just how small? It is a frustrating aspect of disposal planning that these questions have no simple and really satisfactory answers.

The difficulty with specifying a period over which the barriers must guarantee isolation of the waste relates to the gradual decrease in toxicity of the waste over time and to uncertainty about the manner in which the waste might escape. If one imagines that all the waste in a repository could be suddenly set free at some time in the future, and that it would thereupon be readily accessible to the nearby populace, the waste will remain a hazard for more than 100 million years (Figures 2.7 and 2.8) – a period equivalent to geologic time since the early Cretaceous. With a more reasonable guess as to how the escape of waste might occur – say by gradual leakage into ground water moving slowly along a fracture – the time over which functioning of the barriers would be important is more like 10 000 or

100 000 years, since after this the natural environment could probably be trusted to control the greatly reduced radioactivity. Thus a specification of the required isolation time can be set with good reason anywhere between 10^4 and 10^8 years. A possible compromise is to say that a repository should provide assurance of isolation for 10 000 years, with a strong probability that its barriers will continue to function for several hundreds of thousands of years longer. (In talking about times of this magnitude, it is well to keep in mind that recorded history goes back only some 6000 years, and the development of modern industry about 200. Is it plausible to suppose that people with ten millennia more of technological experience behind them will have the same dread of escaping radioactivity that we have today?)

The trouble in attempting to set a permissible release of radioactivity stems from the uncertainty about biological effects of very low-level radiation. If one takes the 'direct proportion, no threshold' hypothesis in all seriousness, then no addition to natural radiation is acceptable; unless waste can be completely contained for all eternity, the exploitation of nuclear energy has imposed an unforgivable burden on future generations. Most opinion is less apocalyptic, granting that radionuclides escaping into the biosphere pose little hazard if the resulting radiation increase is no greater than the observed variations in natural background radiation. This provides a basis for quantitative standards, albeit a rather slippery one. Many feel that it is too restrictive.

By way of illustration, in the United States the agencies charged with the construction and regulation of waste repositories have agreed, after long and often acrimonious debates, on a set of standards to which repositories planned for the next two decades must conform. Engineered barriers (the canister, any sleeve or overpack placed around the canister, and the backfill and seals in the repository) must give reasonable assurance that *no* radionuclides will escape into the surrounding rock during the time when the short-lived fission products (especially ^{90}Sr and ^{137}Cs, both with half-lives of about 30 years) are still abundant enough to be hazardous; this time of containment is specified as at least 300 years, but for some repository environments may be as long as 1000 years, at the discretion of the regulatory agency. As additional assurance that no radionuclides will reach the biosphere during the containment period, the repository must be sited at a place where ground water is known to be moving so slowly that under normal conditions it will take at least 1000 years to move from the repository to the accessible environment (defined as the environment outside a controlled area of 100 km^2, its boundaries nowhere more than 5 km from the repository). For the longer time of isolation, engineering and geologic barriers must be effective enough so that no more than one part in 100 000 of any nuclide remaining in the repository after 1000 years can escape per annum (Figure 3.8). In addition, limits are set on allowable

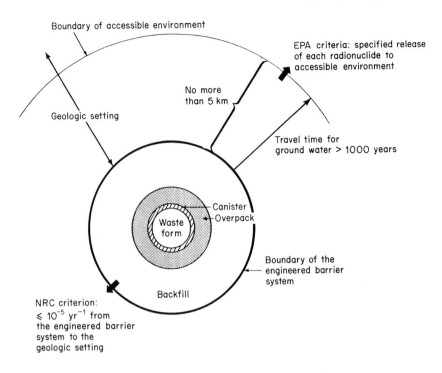

Figure 3.8 Example of standards for radionuclide release, as set by U.S. regulatory agencies. NRC = U.S. Nuclear Regulatory Commission, EPA = U.S. Environmental Protection Agency. (The Analytical Sciences Corporation TR 3336-5, Review of 40 CFR 191 (1984))

radiation exposure to individuals and populations over the first 10 000 years after repository closure. These are examples of the kind of quantitative restrictions that can be established, and that are currently under consideration in many countries. The particular numbers will of course vary from country to country, and may be expected to change with time as more is learned about the chemical and physiological properties of the different radionuclides.

The general strategy of HLW disposal then, is to devise a set of engineered and geologic barriers that restrict the reasonably foreseeable escape of radionuclides into the biosphere within arbitrarily designated limits for at least ten millennia and expectably longer. Both kinds of barrier are essential in controlling radionuclide escape, but because geologic barriers are the focus of attention in this book, problems of the engineered barriers will be less fully discussed.

3.3 ENGINEERED BARRIERS

The planned engineered barriers – the hindrances to movement of ground water and radionuclides that can be constructed or installed and are subject to human control – include the components of the waste package, the backfill, and materials used for sealing shafts, tunnels, and boreholes.

A waste package, one of the units that will be emplaced in holes in the floor or walls of a repository, consists essentially of the waste itself and the metal canister that contains it. Other materials may be included, to help in delaying contact of water with the waste and in slowing the release of radionuclides after contact is made: a matrix (metal or clay) within the canister to fill empty spaces around the waste, or a metal overpack and sleeve around the canister (Figures 3.3 and 3.5). Many designs for the waste package and many kinds of materials have been suggested, and the virtues of different ones are still debated.

The waste form itself may be either spent fuel rods (pellets of UO_2 in zircaloy tubes) or a solid fabricated from reprocessing waste. The most suitable solid form in which to put reprocessing waste has been the subject of long argument. Borosilicate glass is currently the most popular; it has the advantages of ease of preparation at relatively low temperatures (less than $1000°$ C), ability to incorporate large amounts of waste of varying composition into a homogeneous solid, and low solubility in simulated ground water. Glass of this sort is being produced in modest amounts from commercial waste in Europe and Japan, and preparations are under way in the United States to convert military reprocessing waste to borosilicate glass on a large scale. Many feel however, that putting waste into a crystalline form would be preferable. Glass is thermodynamically unstable, they point out, and over the long period of time needed for HLW isolation the glass will at least partially devitrify. This means that tiny crystals will form all through the glass, disrupting its structure and permitting penetration by fluids. Presumably this would make the glass and its contained radionuclides more susceptible to dissolution, but whether the solubility would actually be increased significantly is a matter of dispute.

The crystalline form recommended by most of those who question the long-term stability of glass is a solid designed to simulate the crystals of minerals that are known from their natural occurrence to survive for geologic ages even in harsh environments – minerals like zircon, monazite, and perovskite. Each of the major radionuclides in waste would be incorporated in one of these substances, according to its crystal-chemical properties, and a mixture of these crystalline solids would then constitute the waste form that is to be enclosed in metal canisters. Such combinations, often called 'synroc', are demonstrably less soluble than glass, but whether the slight gain in stability compensates for the greater complexity of fabrication remains an open question. For the next decade or so glass will probably remain the favored waste form, but synroc has strong advocates as a future alternative.

For the canisters that will enclose the waste, there is again no consensus about the best material to use. In Sweden copper is the preferred metal, on the grounds that of all the common metals copper is the slowest to corrode. Extrapolation from lengthy experiments with copper under simulated repository conditions has convinced Swedish workers that canisters with walls 10 cm thick will not completely succumb to corrosion for at least a few hundred thousand years and probably not for well over a million. Despite this impressive demonstration, copper is not widely favored elsewhere, in part because of expense and in part because a repository filled with copper canisters would be too tempting a target for mineral exploration in the future. Steel alloys are a common alternative, in some designs with a surface layer of titanium or a lining of ceramic. In Switzerland current plans call for the use of cast iron, a possibility also under study in the United States; the iron would certainly corrode, but with material as cheap as this the canisters walls can be made thick enough to prevent complete penetration for the necessary containment times. Waste-package designs commonly include a metal overpack around the canister to further slow access of ground water to the waste, and a third metal jacket or 'sleeve' may be used to line the emplacement hole.

Outside the waste package, in the emplacement holes and in the tunnels and other open spaces of the repository, will be another engineered barrier: the material that is used as backfill, intended both to delay movement of ground water towards the canisters and then much later, after the canisters have corroded and waste is exposed to dissolution, to slow the movement of radionuclides away from the waste. The material to be used as backfill is another subject on which final decisions have yet to be made. Presumably it will consist in part of the debris removed in excavating the repository. This would seem particularly suitable for a repository in salt, where the excavated salt returned to the repository would consolidate under pressure and so become part of the massive impermeable salt enclosing the waste. But much may be gained by mixing other substances with the excavated debris, or by using other substances entirely. A commonly mentioned additive or major constituent of backfill is clay, particularly the smectite-rich clay called bentonite, which has the desirable properties of swelling on contact with water and thus plugging all openings, and of serving as an effective sorbent for many radionuclide ions. Other substances might be mixed with the bentonite to control chemical conditions in the filled repository; for example, an admixture of ferrous phosphate would help to keep infiltrating ground water both alkaline and slightly reducing, properties that would slow the rate of corrosive attack on canister metal and could immobilize some escaping radionuclides as insoluble precipitates. Many alternatives are possible for backfill compositions, and recipes can be varied to meet the requirements of different geologic environments.

As a final engineered barrier, the shafts and drillholes penetrating a repository will be sealed with plugs to inhibit access of water from the surface, and other plugs will be placed at intervals along tunnels to control water movement in the fractured rock adjacent to the tunnel openings (Figure 3.9). Presumably the plugging material will be a cement with composition adjusted so as to form a tight seal against rocks of the repository environment. The goal in devising a seal is to ensure that no more water will leak through or around it than would move through a comparable volume of undisturbed rock – a goal that has been approached but not yet completely achieved for all the proposed repository media.

Much research, both in the laboratory and in field demonstrations, has gone into plans for the various engineered barriers. Continuing research is obviously desirable, but it is now largely work on refinements of materials and processes that are already known and need adaptation to particular repository environments.

3.4 GEOLOGIC BARRIERS

The chief geologic barrier to radionuclide movement is the rock in which the waste repository is excavated. Any sort of rock will provide the necessary shielding for radiation, provided only that the repository is placed at least a few hundred meters below the surface, but rock environments differ greatly in their ability to prevent or restrict the movement of ground water and the radionuclides it might carry in solution. A major part of the disposal enterprise therefore, must be a proper choice of the kind of rock and the general geologic environment in which the repository is to be constructed. Finding suitable repository sites is a complex technical problem – and an even more complex political problem, because few communities welcome the prospect of waste disposal nearby, no matter how technically favorable a site may be.

The general characteristics of a geologically favorable site are pretty obvious. The rock at an appropriate depth must be strong enough to maintain an opening, at least for the few years or decades during which the waste is being emplaced. The rock should have low permeability and few fractures or breccia zones through which ground water can move easily. The site should not be in a region where earthquakes are frequent or volcanic activity is likely, or where active erosion might expectably carve to repository depths in the foreseeable future. Places where ore deposits or hydrocarbon accumulations are known or suspected should be avoided, to minimize the chance that human activity might at some time disturb the

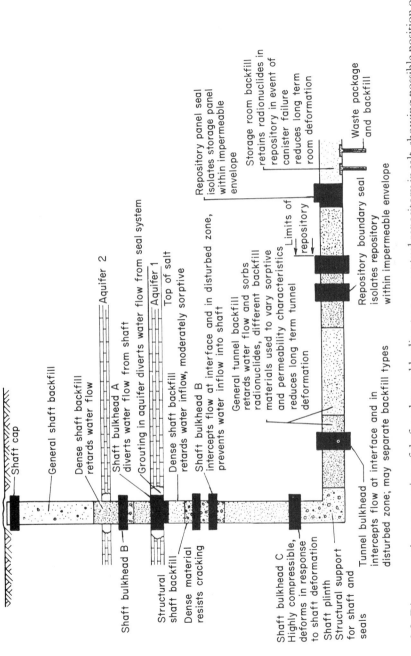

Figure 3.9 Diagrammatic cross-section of shaft and tunnel leading to a conceptual repository in salt, showing possible position of seals ('bulkheads') to control flow of ground water. The seals, consisting of salt blocks or concrete, are made wider than the tunnel openings by inserting them into slots gouged in the walls. This is to ensure sealing of the disturbed zone in the adjacent rock that results from excavating the shaft and tunnels. (D'Appolonia Consulting Engineers Inc.)

The labels in the figure read:

Shaft cap

General shaft backfill

Dense shaft backfill retards water flow

Aquifer 2

Shaft bulkhead A diverts water flow from shaft

Grouting in aquifer diverts water flow from seal system

Aquifer 1

Top of salt

Dense shaft backfill retards water inflow, moderately sorptive

Shaft bulkhead B intercepts flow at interface and in disturbed zone, prevents water inflow into shaft

Shaft bulkhead B

Structural shaft backfill

Dense material resists cracking

Shaft bulkhead C Highly compressible, deforms in response to shaft deformation

Shaft plinth Structural support for shaft and seals

Tunnel bulkhead intercepts flow at interface and in disturbed zone; may separate backfill types

General tunnel backfill retards water flow and sorbs radionuclides, different backfill materials used to vary sorptive and permeability characteristics reduces long term tunnel deformation

Limits of repository

Repository boundary seal isolates repository within impermeable envelope

Repository panel seal isolates storage panel within impermeable envelope

Storage room backfill retains radionuclides in repository in event of canister failure reduces long term room deformation

Waste package and backfill

repository. Ground water at repository depths should be scant and slow-moving, and should have a composition that will not promote corrosion of canisters and dissolution of waste. Advantageous but not essential would be rock in flat layers (to facilitate mining) and rock with high heat conductivity (to help in keeping temperatures low). A dry climate, reasonably flat topography, and accessibility to transportation routes are other evident desiderata. These requirements would certainly eliminate large areas from consideration as repository sites, but are not so restrictive as to make suitable sites formidably difficult to find.

What kinds of rock would be particularly suitable? One that comes to mind immediately is *rock salt*, which certainly satisfies the criterion of scant moving ground water: if any considerable flow had ever existed, the salt would have long since dissolved. Other virtues of salt are its easy minability, its high heat conductivity, and its ability to flow plastically so as to fill openings over periods of decades. Plastic flow in one sense may be counted as a handicap, because mined openings will slowly change dimensions; but a repository can be kept usable during the time required to fill it with waste, and the advantage of a material that will ultimately collapse and fill openings around the canisters outweighs the possible difficulty resulting from slow movement during operation. Another objection to salt is the fact that any water that does accumulate – and all salt contains at least a little water – will be a concentrated brine, highly corrosive to canister metal. Still another is the effect of long-continued radiation on salt, which may liberate chlorine and produce a local oxidizing environment (although recent work indicates that appreciable chlorine could result only from stronger radiation fields than those produced by nuclear waste). Despite these possible cautions, salt remains a most attractive possibility for repository construction. Both bedded salt and salt domes are being seriously considered, especially in the United States and in the Federal Republic of Germany (Figures 3.10 and 3.11).

A second strong candidate for a rock in which to excavate a repository opening is granite or some similar *crystalline rock*. (The term crystalline rock in this context is a misnomer, because salt or limestone or most clay would also consist of crystals, but the name is commonly used for igneous and metamorphic rocks rich in quartz and feldspar, such as granite, gneiss, and some schists.) Crystalline rock is less easy to mine than salt, but it would maintain repository openings indefinitely. In many places the content of ground water is small and its composition noncorrosive. On the other hand, nearly all crystalline rock is cut by a network of fractures and shear zones, some of which can transmit ground water in large amounts. Locating an area of crystalline rock suitable for repository construction is largely a matter of finding a sufficiently large volume of rock at the proper depth in which joints are few and small. A search for such repository sites is

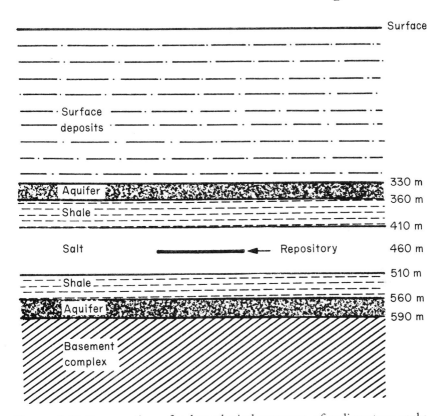

Figure 3.10 Cross-section of a hypothetical sequence of sedimentary rocks, showing the location of a repository in a thick bed of rock salt. Rock sequences of this sort are found in the Permian strata of the south-central United States, in Silurian strata of the northeastern United States and Ontario, and in Devonian strata of western Canada. (U.S. Environmental Protection Agency)

under way in several countries: Sweden, Canada, France, Switzerland, and Japan. In the United States crystalline rock is not being considered at present, but will be a likely choice for a second generation of repositories.

A third kind of rock with many attractive features is *shale*. Obviously not all shales would be suitable, because some varieties are so soft that maintaining an excavation underground would be difficult; but in places where the rock is reasonably competent shale has the virtues of low permeability and high sorbency for ions that move through it. A possible objection to shale is the fact that repository temperatures would have to be kept low, probably under 100° C, because clay minerals lose water and change their character (e.g. from smectite to illite) on heating; the alteration could lead to development of fractures and partial loss of sorbent

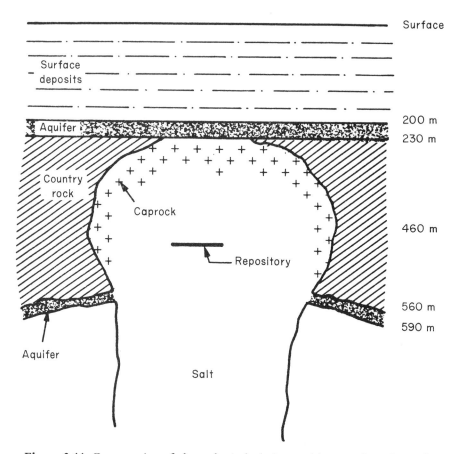

Figure 3.11 Cross-section of a hypothetical salt dome with a repository located in the upper part. Structures of this sort are common along the coast of the Gulf of Mexico and in northern Europe. (U.S. Environmental Protection Agency)

capacity. The difficulties with shale are serious but probably not insuperable, and promising varieties of this rock are being studied as a repository medium in Belgium and Italy.

Another rock under active investigation in the United States is *tuff*, an exceedingly variable material formed by consolidation of volcanic ash. It ranges from loose ash to glassy welded tuff; the former is highly sorbent but could not maintain openings underground, while the latter could maintain openings but has lost much of its sorptive capacity. Some combination of the two should be well suited to repository development, say an excavation in welded or partly welded tuff surrounded by less welded material which would trap any radionuclides that might escape. Tuff seems particularly attractive in the arid south-western part of the United States; in places here a thickness of several hundred meters of tuff above the water table would permit location of a repository in the vadose zone, where it would be free of moving ground water except after rare deluges. Some uncertainty remains about the long-term movement of ground water in such a situation, and about the possible effects of climatic change, but the possibility of locating a repository well above the regional water table has much appeal.

Yet another rock with some possibilities for repository construction is *basalt*. Because basaltic lava commonly occurs in thin flows separated by fragmental material through which water can migrate, and because basalt is often intricately fractured, it seems offhand like a poor choice. In the north-western part of the United States however, some of the units in the huge pile of basalt flows that make up the Columbia River Plateau are thick enough, and in places sufficiently free of joints, to be worth considering. The rock would certainly maintain openings well, except for the joints it is practically impermeable, and associated ground water has the desirable properties of mild alkalinity and low redox potential. Whether these virtues compensate for the probable difficulties in controlling water movement and for the high temperatures that would be encountered during construction (up to 57° C, measured at repository levels) remain uncertain.

Thus for any one of the suggested kinds of rock, its suitability as a disposal medium can be argued back and forth. No rock variety can be designated as ideal in a general sense, and none of the kinds mentioned can be ruled out. Probably any of these rocks, given the right geological circumstances, would be suitable, or could be made suitable, for repository construction. Testing their adequacy at any particular place requires extensive study of the local geologic situation. Examples of such local studies are described in Chapter 6.

3.5 RETRIEVABILITY

Planning for construction of a mined geologic repository is complicated by the problem of retrievability. Should the plans include a provision for retrieving the waste canisters, if this for any reason becomes necessary within a few years or decades after the repository has been filled? Planning is much simpler if such a provision is not needed. The canisters would be emplaced, the holes around them filled, the tunnels backfilled and sealed as each section of the repository is completed, and no care would be needed to ensure the stability of tunnel walls and roofs. This kind of permanent entombment of course, is the professed objective of waste disposal; the

HLW is supposed to be put away in a place where it will never again need to be looked at or disturbed.

Yet plausible reasons can be suggested for thinking that access to the waste canisters ought to be preserved for at least a short while after repository filling. There is always the nagging worry that something might go wrong – that even with all our preliminary research, all our careful estimates as to how the engineered and geologic barriers will perform, somehow radionuclides will escape in unacceptable amounts. Monitoring instruments will be set up around the repository to detect such leaks, and if any are found it ought to be possible to retrieve the canisters quickly and place them in storage on the surface until the fault can be corrected or a better disposal site can be developed elsewhere. A second argument for maintaining the capability of retrieval is that technological advance in the next few decades may find a use for some of the waste – perhaps spent fuel as an energy source, or reprocessing waste as a source of rare metals – and recovery of the canisters would be economically desirable.

Retrieval at best would be a difficult operation. With proper initial planning the difficulties could be reduced – for example, the emplacement of backfill and seals could be delayed, and steps could be taken to keep shafts and tunnels open and in good repair – but such measures add to the cost of repository construction and increase the hazard to operating personnel. Is the retrieval option important enough that it should be a part of basic repository planning, or is the chance of a need for retrieval so remote that it can be disregarded? This is a peculiarly thorny question, to which of course, there is no clearly satisfactory answer.

On the one hand are those who argue that all the redundant safeguards embodied in repository planning make the likelihood of needing retrieval so small it is not worth considering. Besides, if the need is great enough, recovering the waste is always possible, even after a repository has been completely filled and all shafts have been sealed, provided one is willing to spend the necessary time and money on a large-scale mining project. Therefore no provision should be made for early retrieval; waste should be put away permanently, with no thought of ever bringing it to light again. If future generations decide that we were wrong, retrieval is their problem. On the other side are those who feel more keenly the likelihood of human error, who think we must leave open the possibility of correcting our mistakes, or retrieving the waste if necessary within a reasonable time and at a reasonable cost. In the United States the second viewpoint has won favor: regulations require that repositories be so constructed that for at least a few decades waste canisters can be removed from them with an expenditure of time and money not significantly greater than that used in the original waste emplacement. In most other countries the need to make provision for easy retrievability has been considered less important.

By way of summary, strategies for mined geologic disposal of HLW involve setting up multiple barriers to the escape of radionuclides from an excavation deep underground. The barriers are partly geologic, partly engineered. Barriers that can be engineered include the insoluble solid material into which the waste is converted, the metal canister and possible other metal sleeves around the waste, the backfill in tunnels, and the seals anchored in the walls of tunnels and shafts. On the geologic side the chief barrier is the rock in which the repository will be excavated. Several rock varieties seem especially suitable, but for any one kind of rock all the requirements for repository construction will be satisfied only locally, and finding such local favorable areas requires detailed study. Especially important aspects of a possible repository site are the amount, composition, and movement of the local ground water. At a site to be examined in depth, research is needed both on geologic and hydrologic details and on the adaptation of engineered barriers to the particular geologic environment. Provision for easy retrievability of waste canisters is an added complication in repository planning. The many kinds of barriers that make up this scheme for repository siting and construction are to some extent redundant, but the redundancy is needed to compensate for uncertainties in estimates of how the individual barriers will perform.

4

Models of radionuclide release

4.1 MODELS AND SCENARIOS

The disposal of HLW in a mined geologic repository, as detailed in the last chapter, means that several thousand tonnes of hot and poisonous material will be put into an excavation some hundreds of meters under the ground surface. The repository site will be carefully chosen, and a number of artificial barriers will be incorporated to keep the waste from moving. If all goes as expected, no radionuclides will leave the repository for at least several centuries, and thereafter the amount escaping will be very small. For 10 000 years and probably for all time thereafter, the additions to natural levels of radiation near the repository will remain acceptably low. To give substance to phrases like 'very small' and 'acceptably low', regulatory agencies have set numerical limits on the amounts of radionuclides that can be allowed to escape.

This is all very well, but how can we be sure that the regulatory standards will be met? Actual performance of the repository cannot be tested, as would be routine in most engineering projects, because the performance we want to test will be in the distant future. Our only recourse is to set up a model based on an imagined sequence of events by which radionuclides might move from the repository to the biosphere. Each step in the sequence is assigned a number representing the best estimate we can make of the behavior of each radionuclide, using the huge bank of observational and experimental data that have accumulated from research over the past few decades. The rate of escape calculated from the model will be necessarily imprecise, but it should be possible to estimate the degree of uncertainty. Such a model gives us more confidence in repository performance than is gained from vague assurances about escape being 'very small', and it can also help to show where details of the repository location or construction might be improved.

The model can take many forms. Its nature will depend somewhat on how the rate of escape of radionuclides is to be described, for comparison with the permissible rates specified by regulators. The model might be set up for example, to give the amount of each radioisotope that will be liberated into surface environments per year, or the total concentration of radioactive material in the escaping solution, or the dose received by the unlucky individual who happens to stand closest to the point of escape. Models designed to serve these various purposes must all include estimates of the rates at which canisters corrode, rates at which individual radioisotopes dissolve in ground water, and rates at which the different nuclides will travel through the engineered and geologic barriers that impede their progress towards the surface (Figure 4.1). It is these common features of the models that concern us here, rather than details of individual ones.

More elaborate models follow the radionuclides not only to the surface, but also into water that is used by plants and animals and thence into human food supply and into various parts of the body (Figures 4.2 and 4.3). This

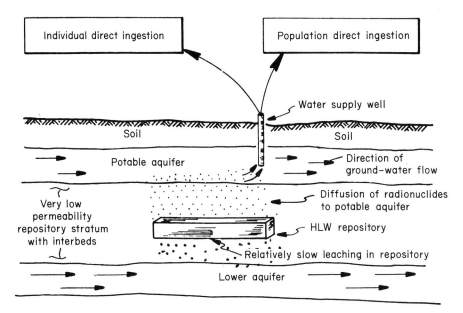

Figure 4.1 Example of a possible scenario for a model of radionuclide release. Steps in the model that would be assigned numbers include the rate of leaching of radionuclides from the repository, their rate of diffusion into the aquifer, and the rate of ground–water movement in the aquifer. The model could give an estimate of either the radiation dose to an individual from ingesting the well water or the dose to an entire population. (U.S. Environmental Protection Agency, Technical Support of Standards for High-level Radioactive Waste Management, EPA 520/4-79-007C (1977))

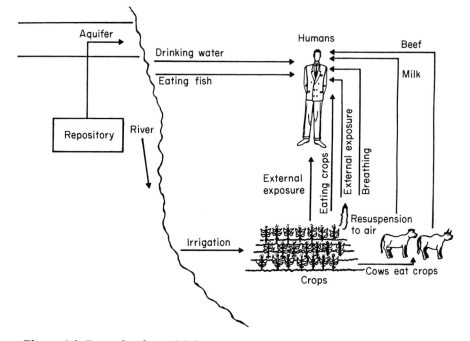

Figure 4.2 Example of a model showing ways by which radioactive material in a contaminated river might cause radiation doses to humans (U.S. Environmental Protection Agency, Population Risks from Disposal of HLW, EPA 520/3-80-006 (1982))

extension of the models is the province of radiochemistry and biophysics, and is beyond our present purposes.

Models can be based on various assumptions about the timing and degree of exposure of waste to ground water. One could suppose for example, that all containers in a repository remain intact for 300 years, that thereafter a few develop pinhole corrosion permitting slight contact of ground water with waste, and that after 1000 years canister disintegration is so complete that contact with waste surfaces is general. This is a reasonable imagined sequence of future events, or scenario, consistent with experimental data on corrosion of some of the metals likely to be used for canisters. It presupposes that everything works according to plan – that all canisters are free from strain and flawlessly welded, that temperatures remain within calculated limits, that ground-water composition does not change, and so on. The scenario is plausible, but perhaps too optimistic in its expectation of ideal behavior.

To make sure that the calculated rate of radionuclide escape is not too optimistically low, a scenario would be preferable that allows for the

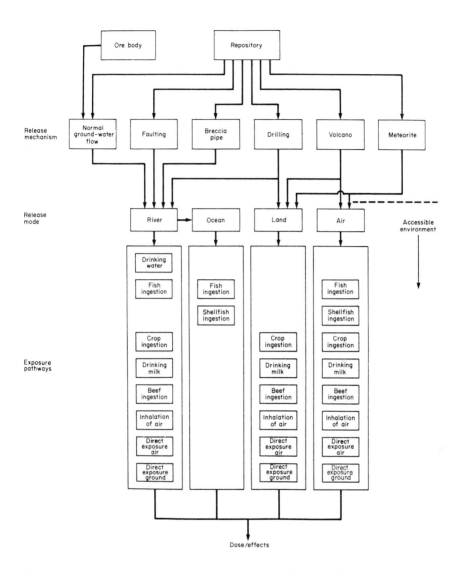

Figure 4.3 The structure of the modeling used by the U.S. Environmental Protection Agency for estimating health effects caused by radioactive releases from waste repositories and from ore bodies. Details of the various release scenarios are not indicated. (U.S. Environmental Protection Agency)

possibility of some human error or carelessness. As an extreme example, one can imagine that the canisters have been so poorly manufactured that they all fail soon after the repository is closed, in other words that waste is directly exposed to ground water from near the beginning. This would be described as a very pessimistic or conservative scenario, in the sense that the rate of escape calculated from it would be almost certainly much larger than the actual rate that could reasonably be expected. Most models from which escape rates have been estimated are deliberately based on conservative scenarios, albeit generally not quite as conservative as this one. In judging estimates of future performance of waste repositories, it is well to keep in mind that those who make the estimates are habituated to using conservative assumptions, so that calculated rates are almost certainly higher than the most probable rates.

To illustrate the considerations that go into model building, we look at some details of a model based on the very conservative or 'worst-case' scenario just described, of canister failure shortly after emplacement and immediate exposure of waste to contact with ground water. This is highly unrealistic, because current study of materials for canisters has shown that several metals and alloys are available which can be depended on to endure in a repository environment for at least a few centuries and probably more than a millennium. It is unrealistic also in that no consideration is given to seals and backfill, which would substantially delay the approach of ground water to the waste surfaces. In effect we are neglecting all engineered barriers except the waste form itself, and focusing attention on the geology. This keeps the model simple, but it is a poor representation of what would actually happen.

The assumption of immediate canister failure means that the whole panoply of radionuclides in waste only a few years out of the reactor will be available to ground-water attack. Thus the very active isotopes ^{90}Sr and ^{137}Cs must figure prominently in our model, whereas with a more realistic scenario we could assume that they have almost entirely decayed by the time ground water touches the waste.

Our model then, envisages slowly moving ground water in contact with waste and carrying materials dissolved from the waste toward the biosphere. Such a model is plausible for a repository in crystalline rock or basalt. It is questionable for one constructed in salt or in tuff above the water-table, since in these environments flowing water would normally be absent, or nearly so. Nevertheless one can imagine circumstances in which water might invade a salt or tuff repository: tectonic movement could open a fissure through which water from an adjacent aquifer might enter, or a radical change in climate could raise the permanent water-table. These unlikely suppositions hardly increase the plausibility of our scenario, but they do give some justification for considering the consequences of early contact of ground water with waste in any sort of repository environment.

The barriers that the model must include are limited to (i) the near insolubility of the waste itself and (ii) the various mechanisms by which the radionuclides are delayed as ground water carries them through the rock on its way to the surface. We need to consider primarily, for each of the radioisotopes, the effects of *solubility* (both of the waste itself and of possible compounds formed en route to the biosphere) and of *sorption* of ions on mineral surfaces (Figure 4.4). To be realistic, the model should also include *dispersion* of ground-water flow, *diffusion* of ions into the rock matrix, and *dilution* by other sources of ground water. Because these items are so dependent on characteristics of particular repository sites, they are omitted from this general discussion. The omission means yet a further increase in the conservatism of the model, for dispersion, diffusion, and dilution would surely aid in reducing radionuclide concentrations.

4.2 SOLUBILITY

Ordinarily we think of the solubility of a simple compound as its concentration in a solution that has stood in contact with the solid compound for a long time, long eough for equilibrium to be established between the rate at which particles of the solid go into solution and the rate at which they reprecipitate on the solid surface. For a substance that is part of a complex solid like nuclear waste the meaning of solubility is less clear. Are the radionuclides actually in contact with the moving water, or are they imprisoned in the glass or crystals that make up the solid waste? Can they dissolve faster than the solid that holds them? Are they uniformly distributed in the waste, or concentrated locally? Is some sort of solubility equilibrium set up, or does the solid disintegrate without any reprecipitation?

Answers to such questions vary with different kinds of waste. In borosilicate glass made from reprocessing waste, radionuclides are presumably distributed pretty much at random through the glass, and can go into solution only as fast as the glass itself dissolves; no equilibrium is possible, because glass is thermodynamically unstable and would not re-form by deposition out of solution. In spent fuel rods the bulk of the waste is crystalline UO_2 and radionuclides may be either enclosed within the crystals or in the interstices between them. Theoretically one would expect (and experiment verifies) that the transuranic elements would be part of the crystal structure, while such fission products as Cs, Sr, and I would be largely excluded from the crystals; the former group would dissolve only as fast as the crystals dissolve, but the latter could go into solution independently. Some of the radionuclide ions from either kind of waste, as they are set free into ground water, would react with the water to form compounds that are less soluble than the waste matrix from which they came; these include chiefly Np, Pu, Am, and Tc, which under the slightly alkaline and reducing conditions typical of most ground water will precipitate as very insoluble oxides (NpO_2, PuO_2, Am_2O_3, Tc_3O_4). Other nuclides, notably isotopes of Sr, Cs, I, and Ra, form no very insoluble compounds with the constituents of ground water, so that their concentrations will be a function of the rates at which the waste dissolves and the ground water moves.

With respect to dissolution in ground water then, the radionuclides in HLW are of two kinds: some that precipitate as insoluble compounds on being liberated from the waste, others whose concentration depends on the rate at which the waste decomposes. For the first group (Np, Pu, Am, Tc, U), their 'solubility' as constituents of waste is simply the solubility of their oxides. Elements of the second group have no defined 'solubility' (except for the solubilities of such compounds as $SrSO_4$ and $RaSO_4$, which are much too high to be effective in controlling radioactive isotopes), and their concentrations could become large if ground water is slow-moving. Concentrations of the first kind are effectively controlled by solubility, concentrations of the second kind are not.

Numerical values of solubility for use in the model may be obtained by calculations with thermodynamic data and by laboratory experiments using compounds that are assumed to be the most insoluble ones under repository conditions. A listing of such values for four different combinations of Eh and pH that might be found in repository ground waters is given in Table 4.1, together with a representative set of values established by one regulatory agency for maximum concentrations permissible in water for ordinary use. The most likely Eh–pH combination to be found in repository water is shown in the first column (slightly alkaline and slightly reducing). Under these conditions, comparison of numbers in the first and last columns shows that low solubility should be an adequate control of concentrations for the elements U, Np, Pu, Am, and Tc. For control of the other elements listed, solubility is manifestly inadequate.

The efficacy of solubility in limiting concentrations of the actinides and technetium however, is not as certain as the numbers in Table 4.1 suggest. For one thing, the numbers are very sensitive to Eh and pH. These elements (except Am) have other oxidation states in which they are more soluble than in the oxides on which the data in the table are based, so that if repository conditions change to make ground water even slightly more acidic or more oxidizing, the control for Tc, U, and Np breaks down. Then the numbers themselves are subject to many qualifications. The accuracy of basic chemical data for some of the elements is uncertain, especially for americium. Numbers in the table refer to crystalline solids;

Table 4.1 Estimates of solubility for important radionuclides: concentration (mg/l) of each radionuclide in equilibrium with its most insoluble compound under the specified Eh and pH at 25° C and 1 atm

	Reducing conditions		Oxidizing conditions		
	Eh = −0.2 V		Eh = +0.2 V		
	pH 9	pH 6	pH 9	pH 6	MPC (mg/l)
Sr	0.6	high	0.6	high	2×10^{-9}
Cs	high	high	high	high	2×10^{-7}
I	high	high	high	high	4×10^{-4}
Tc	10^{-10}	high	high	high	1×10^{-2}
U	10^{-3}	10^{-6}	high	high	5×10^{-3} (for ^{234}U)
Np	10^{-4}	10^{-4}	10^{-2}	10^{-1}	4×10^{-3}
Pu	10^{-5}	10^{-4}	10^{-5}	10^{-3}	8×10^{-5} (for ^{239}Pu)
Am	10^{-8}	10^{-5}	10^{-8}	10^{-5}	1×10^{-6} (for ^{241}Am)
Ra	10^{-3}	10^{-1}	10^{-3}	10^{-1}	3×10^{-8}
Pb	10^{-1}	1	10^{-1}	1	

MPC = maximum permissible concentration in water for ordinary use (USNRC 1976). The designation 'high' means greater than 1 mg/l, a solubility so large that it would not be an effective control of concentration. The table is compiled from many sources. The solubility values are in part derived from experimental data on compounds assumed to be the most insoluble under repository conditions, and in part are calculated from thermochemical data.

for gelatinous hydrated forms of the oxides, such as might well be their state when freshly precipitated, solubilities are known to be much higher. Not indicated in the table is the possible enhancement of solubility by formation of complexes with other constituents of the ground water, or the possibility that the oxides might form first as colloids rather than precipitates, and hence would be capable of moving with ground water in amounts beyond the solubility limit. Despite all these provisos, the figures in Table 4.1 are corroborated by enough laboratory experiments under carefully simulated repository conditions to give some confidence in at least their order of magnitude, and hence in the effectiveness of low solubility in control the movement of the actinides and technetium when ground water is in contact with waste.

4.3 SORPTION

For the elements in Table 4.1 other than technetium and the actinides, low solubility places no limit (or limits too high to matter) on their concentrations. If ground water moves from repository to biosphere with no other control, these elements might well be present in concentrations dangerous to life. The ground water must travel however, through the backfill and then a long distance through solid rock, and during this travel the concentrations of the more soluble nuclides are cut down by sorption on mineral surfaces (and also by dispersion and dilution, which we are neglecting here).

The path followed by ground water is a complicated one, partly through tiny intergranular openings and partly through a network of joints. In a sedimentary rock like sandstone or siltstone the former would predominate, in a crystalline rock the latter (Figure 4.4). In either kind of movement the water would not follow a straight-line path, but would tend to spread out, or disperse, into its surroundings, and would mix with uncontaminated ground water from other sources. With all these complications it is obviously difficult to describe the movement of the water accurately, and even more difficult to describe the behavior of its dissolved ions. Flow through uniform granular material like well-sorted sand can be formulated mathematically in terms of variables like head, permeability, and porosity, and flow along uniform fissures can also be given mathematical expression; but for the nonuniform materials and irregular joint networks in actual rocks the significance of the equations is questionable. It is commonly assumed however, that on a large enough scale – even in a crystalline rock, where flow is chiefly through joints – the movement can be approximated by the equations for flow in granular material. With this simplifying assumption our model lumps together the varying degrees of sorption on different kinds of material – fresh mineral surfaces in the rock matrix, partly weathered surfaces, altered minerals along fractures, secondary minerals in veins – without attempting to distinguish them.

The term 'sorption' covers a multitude of processes, not all of them well understood, ranging from simple adherence of ions to residual charges on a mineral surface to incorporation of ions into the mineral structure by exchange of other ions originally present. If one thinks of a solution flowing slowly through a column of clay in the laboratory, the experimental result is simply that some of the dissolved material is retained in the column. Without inquiring into details, one can say that the clay has *absorbed* the dissolved substances. Or, one can try to picture the mechanism of the retention process, imagining individual ions clinging to the surfaces of clay particles, and hence describe it as *adsorption* on clay-mineral surfaces. Another alternative is to think of some ions from the solution as actively displacing ions that are loosely held in the clay-mineral structure, a process most accurately described as *ion exchange*. Because details of the retention are usually not known, the general term *sorption* is convenient for describing it without reference to particular mechanisms.

Sorption from moving ground water, by whatever means, is generally

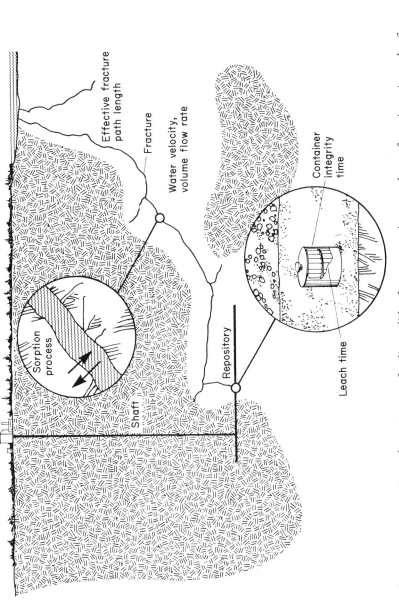

Figure 4.4 Schematic representation of movement of radionuclides from a repository to the surface via an irregular fracture (as might happen in crystalline rock). Various steps in a model of the movement are depicted: the rate of canister corrosion ('container integrity time'), rate of dissolution of waste ('leach time'), rate of movement through the fracture, and sorption of ions en route. (First Annual Report of the Canadian Nuclear Fuel Waste Management Program, Atomic Energy of Canada, Ltd. (1979) eds. I. Boulton and A.R. Gibson)

not permanent, because the first ions sorbed can be displaced by others and put back into solution. Thus for each ion its motion can be visualized as taking place in a series of steps – sorbed–displaced, sorbed–displaced, and so on. The net effect is to retard its movement so that it lags behind the flow of ground water. The amount of lag for a given ion depends on its sorptive properties and on the sorbent properties of the rock through which it is moving. To make quantitative estimates of the degree to which radionuclides are controlled by sorption requires assigning numbers to their retardation.

Efforts to find reliable numbers for this purpose have not been conspicuously successful. The numbers that exist are based on experiments set up to duplicate as nearly as possible the conditions expected along the path of ground-water flow. In the simplest kind of experiment, crushed rock is allowed to stand in contact with simulated ground water containing a radioisotope at low concentration, and the amount of the isotope removed from solution is measured. In other probably more realistic experiments, water percolates down a column or is pumped so that it circulates slowly through crushed rock. In either sort of experiment, when the concentration of radionuclide has ceased to change and a state of equilibrium has presumably been established, a 'sorption coefficient' may be calculated:

$$K_d = \frac{\text{concentration of radionuclide in the solid}}{\text{concentration of radionuclide in the solution}}$$

with units of milliliters/gram. From K_d a 'retardation factor' may be derived:

$$R_f = 1 + K_d \times \frac{\text{density of rock}}{\text{porosity of rock}}$$

which (if dispersion is neglected) is equivalent to the ratio of the rate of ground-water movement to rate of radionuclide movement. Thus an R_f of 10 for an isotope means that it moves 1/10 as fast as the ground water that carries it. The many tables of K_ds and R_fs that have been compiled show a discouraging lack of agreement from one laboratory to another. Some representative values of R_f, given as ranges that include a majority of published values (but not the extremes), are listed in Table 4.2.

The wide ranges in the table are hardly surprising. Sorption depends on many variables, some of them difficult to control experimentally. Temperature, pressure, extent and nature of the solid surface, pH and Eh of the solution, ionic strength, presence of ions that compete for sorption sites with the ion whose K_d is being measured, presence of ligands that can form stable complexes with the ion, all affect K_d measurements. In addition, the

Table 4.2 Retardation factors for important radionuclides in rock materials being considered for repository siting

$$R_f = 1 + K_d \times \frac{\text{density of rock}}{\text{porosity of rock}} = \frac{\text{rate of ground-water movement}}{\text{rate of radionuclide movement}}$$

Element	Granite	Basalt	Tuff (volcanic ash)	Shale or clay	Salt
Sr	20–4 000	50–3 000	100–100 000	100–100 000	10–50
Cs	200–100 000	200–100 000	500–100 000	200–100 000	40–100
I	1	1	1	1	1
Tc	1–40	1–100	1–100	1–40	1–10
U	20–500	50–500	10–400	50–2 000	20–100
Np	10–500	10–200	10–200	40–1 000	10–200
Pu	20–2 000	20–10 000	50–5 000	50–100 000	40–4 000
Am	500–10 000	100–1 000	100–1 000	500–100 000	200–2 000
Ra	50–500	50–500	100–1 000	100–200	20–50
Pb	20–50	20–100	20–100	20–100	1–20

The table is compiled from experimental values reported in many sources. The high figure in each range is an estimate for assumed most common repository conditions (Eh −0.2 to −0.3 V, pH 7 to 9); the low figure is an estimate of the minimum value under less favorable conditions of Eh, pH, or complexing. The numbers for salt do not refer to sorption on salt itself, but on ordinary rock material in the vicinity of a salt repository where the ground water may be fairly concentrated brine.

attainment of equilibrium is hard to verify: in some experiments the sorption ratios appear to increase indefinitely with time, and the reverse process of desorption is often much slower than the original sorption. Published values of K_d are suspect unless there is good evidence that equilibrium was established and all important variables were controlled.

Despite the uncertainties, the experimental work on which Table 4.2 is based provides some useful qualitative generalizations, which are illustrated by figures in the table. Cations are sorbed more strongly than anions, reflecting the fact that residual charges of most mineral surfaces in ordinary rocks are negative; thus the two nuclides in Table 4.2 that would exist dominantly as anions in a repository environment, I^- and TcO_4^-, have much lower retardation numbers than the others. The fact that clay-mineral surfaces are particularly effective as sorbents is reflected in the high numbers for retardation by clay and shale. Sorption is generally less effective for radionuclides in brines than for those in dilute ground water, because ions of the brine compete with the radionuclide ions for sorption sites; hence the numbers for retardation by rocks near a salt repository (last column) are conspicuously lower than those for other rocks. Thus the table gives at least a rough indication of the probable behavior of different radioisotopes in various repository environments.

We can expect then, that concentrations of the actinide elements will be strongly limited by both solubility and sorption – first by the low solubility of their oxides, then by the retardation of any cationic forms that do get into solution. Among the elements not controlled by solubility, Cs and Sr have such high retardation factors that acceptably low concentrations seem assured as ground water traverses almost any kind of rock. For Ra and Pb the numbers are smaller and adequate control is less certain. Iodine remains completely uncontrolled. The low numbers for Tc indicate that sorption can contribute little to its control, and for Np the adequacy of control is questionable; for both of these nuclides the major control is the insolubility of their oxides, and this is effective only if ground water remains reducing and slightly alkaline.

From this summary, what conclusion is reasonable about the ability of our model to ensure against hazardous releases of radionuclides to the environment in the event of early canister failure? Acceptably low concentrations of Cs, Sr, U, Pu, and Am seem assured by low solubilities or high retardations or both, in almost any kind of repository environment. Np and Tc can be included in this group provided that Eh and pH conditions remain within narrow limits; conditions that stray from these limits would be most likely in the vicinity of repositories constructed in salt or in the vadose zone in tuff, but these are precisely the environments where moving ground water should be absent, or nearly so. Adequate control of ^{226}Ra and ^{210}Pb is somewhat questionable, but these isotopes (resulting from decay of the actinide elements) would become important only after some 10 000 years, and would be in large degree limited by sorption. Except for iodine then, the long-lived constituents of HLW will be effectively kept out of surface environments, and most of the iodine, according to present plans, will be removed from the waste before burial. Thus a rudimentary model with only two numbers – for solubility and retardation by sorption – already gives some assurance that the amount of radioactive material escaping from a repository during the next 10 000 or 100 000 years will be within or close to the limits permitted by regulatory agencies.

A more complete model would include other numbers, for example numbers relating to ground water – its quantity, rate of movement, distance of travel to the biosphere. Such numbers are not easy to obtain, but can be estimated from geologic observations at the surface and standard hydrologic measurements in drillholes. Then there would be numbers for dispersion of ground-water flow, diffusion of ions into the rock matrix, and dilution by uncontaminated ground water. Finally the model would include rates at which different radionuclides move along pathways from their point of emergence at the surface into their possible human victims (Figures 4.2 and 4.3). These numbers of course, would all add to the assurance that the escape of radioactivity into human environments would not present an appreciable hazard.

This cheerful conclusion, be it remembered, is derived from a model that ignores engineered barriers and that is based on an absurdly conservative scenario in which waste is exposed to ground water immediately after repository closure. But like all such conclusions in the complex world of waste disposal, this one must be taken with a pinch of salt – as will be evident in the next chapter.

4.4 ALTERNATIVE MODELS

Rather than ignore engineered barriers, as we have done so far, we might just as reasonably construct a model in which engineered barriers have a central role and geologic barriers are downplayed. This kind of model has been favored in Sweden, where current plans call for putting thick-walled copper canisters in a crystalline-rock repository that will be backfilled with bentonite or a bentonite–sand mixture (Figures 3.6 and 3.7). Experimentally determined corrosion rates indicate, by reasonable extrapolation, that the canisters will completely contain the waste for at least a few hundred thousand years and probably more than a million; if something should go amiss with a few canisters, the bentonite will retard both movement of ground water to the waste and movement of radionuclides away from it. Geologic barriers are irrelevant here, except that the composition of ground water must not change to become corrosive.

In this model no radionuclides escape at all, so calculations of concentration are pointless. But the Swedish model, to make doubly sure, considers scenarios in which canisters or bentonite or both are assumed to fail at various times after repository closure. The models for these scenarios are similar to the one we have described, depending chiefly on geologic barriers but tailored specifically to conditions in Swedish bedrock. The Swedish work includes comprehensive theoretical and experimental studies of solubility and sorption under repository conditions, and in addition makes an attempt to assign numbers to ground-water dispersion and to diffusion of ions into the rock matrix along fissures. A general conclusion from the Swedish work, as from studies elsewhere, is that adequate control of radionuclide escape is possible with either engineered or geologic barriers acting separately, but is more certain when both are present.

This discussion by no means exhausts the catalog of models that have been constructed, nor does it follow any of the models to the ultimate stage where actual doses to individuals or populations are calculated. But it suggests the general pattern of model building, and the kinds of assumed scenarios on which models are based. It should be emphasized once again that the numbers used throughout are chosen to be conservative, so that the calculated amounts of escaping radioactive material are in all probability exaggerated. Even with the conservative assumptions, plausible models lead uniformly to the conclusion that amounts of radionuclides likely to escape from a well-sited and well-engineered repository can be kept within acceptable limits.

Critique of disposal models

5.1 QUESTIONS FROM A SKEPTIC

Model building, as described in chapter 4, gives comforting assurance that geologic disposal can indeed accomplish its purpose of isolating the radionuclides in HLW. But a skeptic may not be entirely convinced: he or she can ask many questions about details of the models, and until these questions are faced he or she can rightly claim that the models seem like mathematical abstractions with little relevance to the real world of rocks and underground water. In this chapter we consider some of the more embarrassing objections that a confirmed skeptic can raise to the assumptions implicit in our models.

One question relates to the numbers used to express solubilities and retardations (Tables 4.1 and 4.2). Admittedly much uncertainty is attached to some of these values; isn't it possible that the uncertainties are large enough to invalidate conclusions drawn from the numbers? The numbers depend on an assumption of reducing conditions in a repository; what if conditions are not reducing? After all, the repository will be full of air when it is sealed; couldn't oxidizing conditions persist long enough to corrode the canisters, expose the waste, and make some of the radionuclides soluble? What about the heat that will be generated by the waste for hundreds or thousands of years: couldn't long, continuously high temperatures promote unexpected reactions between ground water, backfill, and surrounding rock? Wouldn't the heating open fractures in the rock, or set up convection currents in the ground water, that would cast doubt on the models' usual assumption of uniform, very slow ground-water movement? What about the long-term effects of intense radiation: couldn't minerals be altered, or water decomposed, so as to change chemical conditions in the repository? What if the climate changes, with consequent variation in the amount and/or composition of ground water? What if a major geologic event occurs – deep erosion, a large earthquake, a meteor impact? What if some of our descendants, seeking minerals or a source of potable water, inadvertently penetrate a repository with drillholes? With a little imagination such questions can be multiplied indefinitely. They are often asked, both by laymen and by those in the technical community who remain unconvinced that the underground disposal of HLW is completely safe.

The questions deserve answers, even though some of them seem trivial or outlandish. With geologic disposal, we shall be committed to trying a radically new technology – putting large quantities of highly dangerous material underground, and leaving it there for times unprecedented in human experience. For such a venture, careful consideration is needed of all the things that can possibly go awry. To judge the effectiveness of planned disposal procedures our only recourse is the building of models, and all details of the models need scrutiny with an appropriate degree of healthy skepticism.

We look now at some of these details, at the doubts they can raise in the mind of a skeptic, and at some of the possible answers.

5.2 UNCERTAINTY IN SOLUBILITIES AND RETARDATION FACTORS

The manifold reasons for ambiguity in the numbers for solubility have been mentioned before – slow attainment of equilibrium, small size of crystals, changes of oxidation state, presence of complex-forming ligands in the ground water, formation of mobile colloids rather than precipitates. These possibilities obviously demand caution in the use of solubility values that are calculated from tables of thermochemical data for equilibrium conditions. Freshly precipitated $Pu(OH)_4$ for example, has a solubility several orders of magnitude higher than that calculated for crystalline PuO_2, it may remain suspended as a colloid rather than precipitating, its solubility may be increased if ground water contains complex-forming ions like CO_3^{2-} and HPO_4^{2-}, it becomes more soluble if the redox potential strays from the range that is common in repository environments. With all these possibilities, our skeptic seems justified in wondering whether the tabulated values for solubility have any meaning at all. In defense of the numbers, one can point out that they are corroborated by many experiments set up to duplicate conditions expected in a repository, that repository sites are chosen where ground water has appropriate values of Eh and pH and a minimum of complex-forming ions, and that approach to equilibrium is probable because of the long periods of time available and the somewhat elevated temperatures of a repository environment. One can note also that a rule of model-building is to choose conservative values for all variables, thus solubilities at the high end of a given range.

More vulnerable are the numbers pertaining to sorption. The uncertainty in retardation values is amply illustrated by the wide ranges in Table 4.2. The numbers can be attacked also on procedural grounds: the values of K_d are sensitive to the nature of the surface with which ground water comes in contact, and there is serious question as to whether the clean, fresh surfaces of the crushed rock commonly used in laboratory work have any resemblance at all to the partly weathered, altered, or coated surfaces that ground water will encounter as it moves away from a repository. Doubts of this sort have arisen from field experiments in which ions are injected into one well and monitored in adjacent wells, experiments that have often given retardation factors smaller than those measured in the laboratory. The relevance of measured K_ds to predictions about ion migration is so questionable that some have advocated abandoning them altogether as indicators of radionuclide behavior. So extreme a view seems hardly justified, in light of the obvious fact that cations do show sorption in amounts that can be roughly correlated with their nature and with the nature of the rocks through which they travel, but the specific numbers must be used with appropriate caution. Here, as for solubility, the fact that only conservative values of retardation numbers are chosen for modeling gives some assurance that the models are dependable.

A skeptic's worry that uncertainty in the numbers used for solubility and retardation may give false assurance of low release rates can also be assuaged by noting that further conservatism is assured in many models by neglect of the substantial decrease in radionuclide concentrations that may result from dispersion and dilution during ground-water transport.

5.3 INITIAL OXIDIZING CONDITIONS

Our skeptic is certainly right in noting that conditions in a repository immediately after it has been filled with waste and sealed may be potentially destructive to waste packages, because of oxygen and carbon dioxide in the air that will remain in all interstices of the backfilled shafts and tunnels. The canisters will start to corrode, and if any waste is exposed some of its radionuclides would surely be susceptible to rapid solution. The harm that this might do will depend of course, on the amount of air present. An estimate of the amount can be made very simply from the geometry of the planned repository, and it turns out to be so small that no more than superficial corrosion would be expected. The oxygen and carbon dioxide will soon be exhausted by the corrosion reaction and by reactions with minerals in the rock and backfill, and as ground water penetrates the interstices it will re-establish the conditions of low redox potential and slight alkalinity that existed before the repository was excavated. Presumably oxidizing conditions might persist longer in a salt repository, where reducing materials are less abundant in the rock, but the products of corrosion and possible dissolution would remain immobile unless the repository is mechanically disturbed.

Thus a skeptic's worries about effects of trapped air can be readily disposed of, but he or she could counter by suggesting that air might at some later date penetrate from the surface and confound the calculations from our model. The seals in shafts and tunnels are designed to prevent such a calamity, and only extreme pessimism could lead to doubts about their capacity to do so. Even if the sealing is not ideal, movement of air into the repository would be so slow that it could not significantly affect the benign chemical conditions maintained by the ground water, the backfill, and the enclosing rock.

5.4 EFFECTS OF HEAT

The heat generated by radioactive decay in a sealed repository will raise rock temperatures to a maximum within a few decades after repository closure and then gradually subside. The temperature distribution at any time can be readily calculated as a function of the kind and age of the waste, its concentration of heat-producing isotopes, the spacing of canisters in the repository, and the thermal conductivity of the rock (Figures 5.1 and 5.2). Because these variables are adjustable when the repository is being planned,

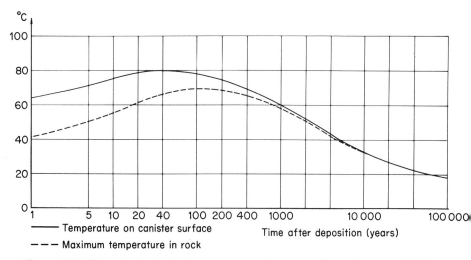

—— Temperature on canister surface
– – – Maximum temperature in rock

Time after deposition (years)

Figure 5.1 The calculated rise in temperature in a repository for spent fuel in crystalline rock in Sweden. The amount and age of fuel in a canister are adjusted to keep the maximum temperature no higher than 80°C. The maximum at the canister surface is reached 40 years after closure of the repository. The maximum in the rock, at the edge of the backfill around the canister, is not quite as high and is reached several decades later. (Svensk Kärnbränslehantering AB, Final Storage of Spent Nuclear Fuel, KBS-3, I General (1983))

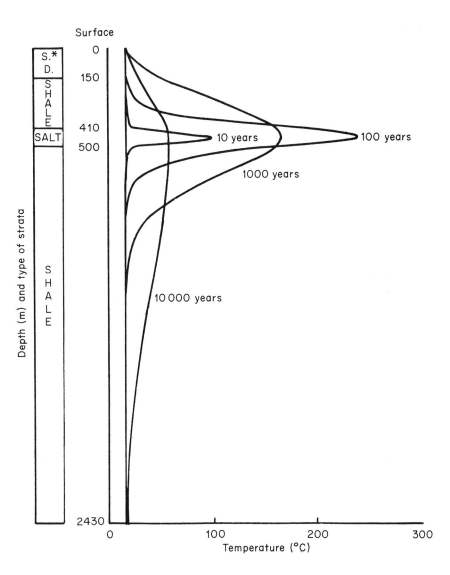

Figure 5.2 The calculated vertical temperature distribution in rocks near a repository for reprocessing waste in a salt bed at a depth of 460 m, with shale layers above and below. The temperature rises to a maximum at the level of the repository in 100 years; thereafter the temperature falls but a longer vertical section is affected. (U.S. Environmental Protection Agency, Technical Support Document B, 520/4-79-007B (1979)) (Based on 10-year-old reprocessed waste with planar heat density of 150kW/acre at burial time.)
★Sedimentary deposits.

anticipated temperatures can be kept below any specified value. (Low temperatures come only with increasing the cost of the repository of course, because a larger number and wider spacing of canisters containing low radioisotope concentrations would be required.) Present plans for repositories in Canada and Europe set a limit at 100° C or less, and in the United States a limit of 250° C. Even at these modest figures long-continuous heating can affect rocks in ways that are not easily predictable. A skeptic is quite justified in questioning whether, or how well, our models take account of such effects.

One consequence of heating is the dehydration of water-bearing minerals, for example clay minerals in shale and complex evaporite minerals that are often found as impurities in salt. Loss of water may change the rock structure, forming minute open spaces and possibly small fractures, thereby increasing permeability and permitting faster ground-water flow. The effect is hard to predict, since it depends on time, temperature rise, and amount and stability of the minerals concerned. If such minerals are abundant, a model of repository behavior must include provisions for keeping temperatures well below the level at which they decompose.

Heating may also cause mineral changes, whether or not the changes involve dehydration. For example, the smectite minerals that are often abundant in shales change on moderate heating to illite. This change makes the rock more porous and less sorptive, alterations that impair its ability to retard moving radionuclides. How serious the effect is for any specific shale can only be determined by experiment. The change can be prevented or greatly slowed by keeping rock temperatures below 100° C.

Since most planned repositories are located in rock that is saturated with ground water, a possible effect of heat would seem to be the setting up of convection currents. Models generally assume uniform motion of water past or through a repository, with only minor disturbance from repository components. If heating is sufficient to cause a substantial vertical rise in the water directly over the repository however, convectional circulation might be established and radionuclides dissolved from the waste could thus find a short cut to the earth's surface. Calculations show however, that this will generally not happen; at the planned low temperatures, the forces leading to convection would be too small to counteract the flows of ground water that are usually encountered. This conclusion obviously would not hold if the flow is abnormally slow or the temperature abnormally high.

Finally, heat would certainly cause expansion of the rock near a repository, and the expansion could conceivably cause open joints to close or new cracks to open. This effect depends intimately on the nature of the rock at any particular repository site, and is seldom predictable from generalized rock properties. The only way it can be incorporated in models is by using *in situ* experiments with artificial heaters, after the site has been opened with a shaft and one or two exploratory tunnels. Experiments of this sort that have been conducted in underground laboratories suggest that the planned temperatures are not high enough to cause harmful changes in rock structure in most repository environments, but this conclusion will need checking as each site is opened.

In general, experiments and calculations indicate that the effects of heating by the HLW in a repository are not likely to invalidate current models of radionuclide release, provided that temperatures remain low. Just how low is an open question: certainly the temperatures proposed by Europeans, 100° C or below, seem sufficiently modest, but the higher temperatures being considered in the United States (up to 250° C) are more questionable.

5.5 EFFECTS OF RADIATION

Potentially more serious than the effects of high temperature, and more difficult to predict, are possible changes produced by the intense radiation field around the waste. Experiments have shown that most kinds of rock under consideration as repository media are not notably changed by radiation; salt is slowly decomposed with liberation of traces of chlorine, but only by radiation more intense than that expected in a repository. The commonly suggested backfill material, bentonite clay, has also been shown to be little affected by radiation at repository intensities. Radiation effects have not been thoroughly studied at high temperatures however, and a skeptic could reasonably deplore the inadequacy of such research.

The change most likely to cause trouble is the dissociation, or radiolysis, of water. Radiolysis is a complex reaction that produces H_2 and several oxidizing species, including O_2 and various peroxide ions in solution. In a closed system the hydrogen would slowly react with the oxidizing materials to re-form H_2O, but because H_2 diffuses through the surrounding rock faster than the other substances, a strongly oxidizing solution may be left near the canisters. If the water contains nitrogen, another product of radiolysis may be nitric acid. Thus a potent acid oxidizing solution may accumulate near canister surfaces, and corrosion could be greatly accelerated. How serious a danger this might pose is hard to say, because it obviously depends on such local variables as rock permeability, ground-water composition, and rate of water movement. If the oxidizing solution is swept away from the canisters by ground-water flow, one would expect both the acid and the oxidizing solutes to be neutralized within a short distance by reaction with constituents of the backfill and rock. If any radionuclides have dissolved, most of them would presumably precipitate at the interface between the oxidizing solution and the normal reducing ground water in the surroundings. In effect, a substantial volume of acid oxidizing solution with dissolved radionuclides might form in the immediate vicinity of the waste, but the nuclides would largely precipitate, or be sorbed, as ground water carries them into the unoxidized rock. Thus one can argue that the overall effect of radiation on waste isolation is probably small, but the argument depends on so many assumptions that it is not altogether convincing.

A skeptic trying to build a case for neglect of important factors in current models would do well to seize on the lack of precise data about the effects of radiolysis. Those who choose to defend the models should emphasize the need for additional research on the interaction of radiation, especially at somewhat elevated temperatures, with the many combinations of substances that may be found in a waste repository.

5.6 GEOLOGIC AND METEOROLOGIC EVENTS

What if a violent earthquake causes the roof of a repository to collapse, or opens a fissure that provides an escape route for ground water containing radionuclides? What if a new volcano begins its activity with an explosive eruption directly under a repository? What if torrential rains cause stream patterns to change, so that a huge gully is opened to repository depths? What if slow tectonic movement causes the land near a repository to be raised and tilted, so that erosion is greatly accelerated? What if renewed glaciation should bring an ice cap over a repository, or cause deep erosion from melt-water? What if a large meteorite should strike the surface at a repository site? This is only a sample of the 'what if' scenarios that a dedicated skeptic would claim should be included in our models of radionuclide release.

It is true of course, that all of these suggested events are possible. Disasters of this sort have happened in the past, as the geologic record clearly shows, and they can certainly be expected in the future. They could indeed raise havoc with a repository, and might spread radioactive debris far and wide. The argument against taking them seriously must depend chiefly on their extreme rarity. From the frequency of such events over geologic time, one can give semi-quantitative expression to their probability in any small area of tectonically stable parts of the continental crust. The resulting numbers are in the range 10^{-7} to 10^{-11} per annum, meaning that an event of this sort can be expected no more often than once in ten million years. Surely in ordinary life we have no qualms about accepting risks of catastrophic events that are far more probable than this.

A few of these imagined calamities need further discussion. While earthquakes that might cause a major break directly through the rocks of a repository are too unlikely for serious consideration, seismic events of lesser magnitude must certainly be expected. Strong earthquakes have occurred even in areas commonly thought to be tectonically stable, and it is reasonable to inquire what damage a repository might sustain from violent shaking. The best evidence comes from records of the observed behavior of underground installations such as mines and tunnels, and this evidence shows clearly that deep subsurface workings are only minimally affected. Dislodging of loose material and some spalling of rock walls are reported, but little structural damage. A repository should certainly be located in a region of low seismic risk to minimize possible effects on surface installations, but the chance of serious seismic damage to the repository itself seems very small.

Similarly for volcanoes: the opening of a vent directly through a repository would surely be a major catastrophe, but other manifestations of volcanic activity – lava flows, ash falls, explosive release of gas – would have little influence on a nearby repository.

Of more concern is the possible effect of a drastic alteration in weather patterns. We have good evidence for marked changes in climate over the past 10^5 years, and similar changes are expected in the future. The cyclic record of Pleistocene glaciations in fact, makes the reappearance of extensive ice sheets very probable sometime in the next ten millennia. Glacial scour is unlikely to be deep enough to affect a repository directly, and extensive erosion by melt-water seems improbable at a well-sited repository; but possible changes in amounts and rate of movement of ground water caused by a thick overlying mass of ice would be well worth further study. Southward from the probable limit of future glaciation the principal expectable climatic change would be an increase in rainfall, but the increase is not likely to be damaging. Pleistocene climates in the American southwest for example, brought maximum annual rainfall to no more than three times the present amount, a change that would not seriously affect a repository deep underground.

Thus the concerns of a skeptic about failure to give imagined geologic and meteorologic events a prominent place in models of radionuclide release seem largely unwarranted. For many such events the proper choice of repository sites can minimize the chance of jeopardizing waste isolation, and for others the extreme improbability of occurrence seems an adequate safeguard.

5.7 HUMAN ERROR AND EQUIPMENT FAILURE

In most release scenarios the engineered barriers of a repository are assumed to be well constructed and to play their assigned roles without a flaw. The waste form retains its composition and structure, the backfill is evenly distributed and uniform in composition, each canister is tightly welded and corrodes evenly, and ground–water movement is controlled by grouting of cracks, nearly impermeable cement seals, and unchanging structures of rock and backfill. So perfect a performance is hardly realistic, our skeptic might say with good reason, because all parts of the scenario involve human activity and human judgment, plus the functioning of large-scale equipment. Somewhere, sometime, human error and equipment malfunction will lead to defects in the operation. Perhaps the metal of a canister is flawed, or its cover was carelessly welded. Perhaps the backfill was carelessly emplaced, and open spaces remain around canisters. Perhaps the grouting of a major joint fails, or the faulty seal of a shaft permits entry of water from the surface. Perhaps a canister was dropped, and its metal cracked, while it was being lowered down the shaft.

How greatly would such mishaps and errors compromise the isolation of waste? Each one by itself, probably not very seriously. A combination of errors would be more worrisome, and if this were likely our skeptic might justifiably maintain that human failings could invalidate release models. The best answer to him or her goes back to improbability, as for volcanic activity and meteorite impact: a damaging combination of errors is always possible, but extremely unlikely. The disposal system is designed with enough redundancy and enough conservatism to compensate for a large number of mistakes. Human error cannot be entirely written off as a possible contributor to repository failure, but the chance of a major increase in radionuclide escape seems small.

5.8 HUMAN INTRUSION

The event most likely to cause a significant release of radionuclides at some time in the future, and the event most troublesome to include in our models, is the drilling of a borehole into the repository. The drilling may be deliberate, in the hope of exploiting some constituent of the buried material for monetary gain. Or one can imagine a scenario in which all record of the repository has been lost and the drilling is merely part of a random search for metallic ore or oil or potable water. Deliberate drilling is especially probable if the waste consists of spent fuel, because its contained fissile material (^{235}U and ^{239}Pu) may appear to entrepreneurs of a future generation as a valuable source of energy. Or some other part of the waste package may seem economically attractive: thus a major objection to the Swedish plan for waste disposal is the likelihood of future excavation to recover the copper of the massive canisters. Presumably those who drill into a repository knowingly will do so with their eyes open, and will take suitable precautions, but the unfortunates who penetrate buried waste or contaminated ground water accidentally may bring large quantities of radioactive material into the biosphere and may be quite unaware of what they are doing.

This eventuality is peculiarly difficult to guard against, and to include in release models, because we cannot guess the state of technological advancement at a time in the distant future. One can imagine a happy time when ways have been found to prevent or quickly cure cellular damage from radiation, so that an accidental encounter with radioactivity would be of no consequence; or one can suppose that future exploration will be done with drills that have automatic monitors for radioactive material, so a driller would be alerted before any damage is done. But it seems equally conceivable that civilization will regress rather than progress, and that primitive people of the future might accidentally encounter radioactive material and be unaware of its dangers. Much thought has been given to ways of marking repository sites permanently and maintaining records of their existence, so that chances of accidental intrusion into a repository are minimized, whatever the state of future civilization may be.

This is hardly a complete survey of the ways a skeptical imagination could suggest by which our models of repository performance might fail, but it is enough to indicate the kinds of 'what if' objections that are often raised and the kinds of answers that can be given. The best answer to some of a skeptic's questions is to call on simple probability: if a postulated event or circumstance is so unlikely that the risk involved is far less than risks we ordinarily accept without question, the objection loses much of its force. Other questions pin-point properties of waste or rocks or ground water about which technical knowledge is still inadequate, and on which additional research would be helpful. The amount of research already accomplished to explore points raised by skeptics is impressive, but uncertainty remains about some important matters, for example the effects of radiolysis on ground water in different kinds of rock, especially when combined with high temperature, and the long-term effects of heating in the presence of brine on the properties of material to be used for backfill and sealing. There is widespread agreement among scientists and engineers that objections to the accepted models of repository behavior can be satisfactorily answered, but there is equal agreement that research should continue on the few points that remain in doubt.

6

The geology of repository sites

6.1 GENERAL REQUIREMENTS

A recurrent theme in this discussion of HLW disposal has been the need for carefully chosen repository sites. How is such careful choosing to be accomplished? In broad areas where the underlying geology seems generally suitable, how does one pick particular sites for detailed study or for actually starting repository excavation? What information is needed, and how can it be obtained with a reasonable expenditure of time and money? These are basic questions that face a geologist when called on for advice about disposal of high-level nuclear waste.

The general attributes of an adequate site are obvious enough, and have been noted before: a rock at a depth of 300 to 1500 meters that will maintain an opening for at least a few decades, that has low permeability and few large fractures, and that is capable of sorbing dissolved material; ground water that is slow-moving, dilute, and of noncorrosive composition; absence of metallic ore, hydrocarbon resources, or other materials likely to be targets for exploitation in the foreseeable future; an area where tectonic disturbance, volcanic activity, and deep erosion are unlikely; and preferably a place where the topography is suitable for a large engineering operation. In the many general areas that would satisfy these requirements, how does one narrow the search to the particular sites where geologic indications seem especially favorable?

The question is best answered by looking at actual examples of studies currently in progress for locating repository sites. For this purpose we choose three such studies, which illustrate the problems of three different geologic media: crystalline rock in Sweden, bedded salt in Texas, and tuff in Nevada. No actual site has as yet been definitely established in any of the three, but several possibilities have been or are being carefully scrutinized.

6.2 CRYSTALLINE ROCK IN SWEDEN

The bedrock underlying most of Sweden consists of granite and various kinds of metamorphic rock, so that Swedish geologists have little choice – the disposal site they seek must somehow be located in crystalline rock. They have risen admirably to the challenge, and their work on the possibilities of crystalline rock as a repository medium is a model for similar studies elsewhere.

The Swedish plan, as noted earlier, relies more heavily than other disposal plans on engineered barriers to isolate waste from the biosphere. HLW will be enclosed in thick-walled copper canisters placed 500 m or more below the ground surface, surrounded and overlain by bentonite, and spaced so that rock temperatures will not rise above 80° C. (Figures 3.6 and 3.7). Extrapolation from experimental work indicates that such canisters in contact with ground water like that found in Swedish bedrock will maintain their integrity for at least a million years, and the biosphere will be completely protected from release of radionuclides until most of them have decayed to harmless levels. With engineered barriers so carefully designed, geologic barriers seem almost superfluous. One could put the copper canisters almost anywhere underground, provided the ground water is not grossly abnormal in amount or composition, and the waste would be safely isolated.

But the Swedes are taking no chances. They know that accidents are always possible – the canister metal may be defective or inadequately welded, fault movement could rupture some of the canisters, ground-water composition could change – and they plan to provide effective geologic barriers as a backup to the seemingly impregnable canisters. For this purpose Swedish geologists have made elaborate studies of many possible repository sites, using a variety of techniques to characterize large areas of crystalline rock in a degree of detail not equaled in other countries. The exploration so far is based entirely on surface mapping, geophysical measurements, and data from drillholes. The geologists are aware of course, that additional data from exploratory shafts and tunnels will be needed when the time comes for actual site selection.

Bedrock in Sweden is part of the Baltic Shield, a large area of Precambrian rock in northern Europe that has been tectonically quiescent except for minor uplift and subsidence for most of the past 600 million years. It is cut into large blocks by major faults, and within each block is a network of minor faults, joints, and shear zones. Most of the faults have been inactive since the Precambrian, but a few of the larger ones show evidence of Phanerozoic movement – as if the blocks have been subjected to minor jostling, while their interiors remained motionless. Movement of ground water in the crystalline rock is almost entirely through the fractures and shear zones, and the search for repository sites is largely an effort to find places where such potential channels of movement are few.

The search is difficult because bedrock in Sweden is largely concealed by glacial debris and abundant vegetation. Fortunately the major faults and shear zones betray themselves by subtle effects on the topography: since they are zones of weakness and of ground-water movement, they are commonly followed by small valleys and belts of vegetation. On air photographs such zones appear as vaguely defined lineaments. The first step in the exploration then, is the plotting of lineaments, aided by mapping of bedrock structures wherever they are exposed and by extensive use of geophysical measurements. The initial mapping showed major lineaments over large areas (Figure 6.1), then was narrowed to progressively smaller areas selected as places where lineaments are scarce. Ultimately several areas

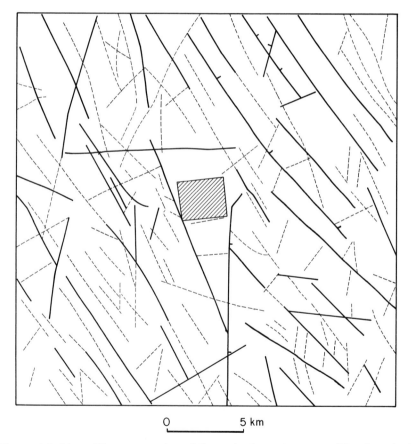

0   5 km

Figure 6.1 Map of lineaments plotted from air photos in an area (Kamlunge) of about 650 km² in northern Sweden. The shaded rectangle shows the area chosen for more detailed study. (Svensk Kärnbränslehantering AB, Final Storage of Spent Nuclear Fuel, KBS-3 IV Safety (1983))

of a few tens of square kilometers were picked for more detailed study. The final choice was often limited by the reluctance of some property owners to have their land considered for waste disposal – in Sweden, as elsewhere, the fear of radioactivity is widespread.

Each site selected for thorough investigation was mapped in as much detail as the scanty outcrops permitted, and drilling sites were located to give the maximum information underground (Figure 6.2). Many holes were drilled, some to depths greater than 600 m (the planned depth of a potential repository), and some at an angle so as to intersect prominent vertical or near-vertical joints and shear zones. By logging the drillholes and examining the cores, fractures could be precisely located and a three-dimensional model of the fracture system could be constructed (Figure 6.3). A question that commonly arises in such a program of mapping and

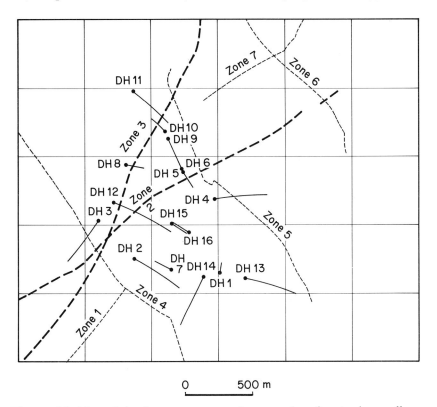

Figure 6.2 Map of the fracture zones at the ground surface in the small area (30 km²) that is shown by shading in Figure 6.1. Heavy dashed lines are fracture zones 5 to 15 m wide, light dashed lines those less than 5 m wide. Light solid lines are horizontal projections of inclined drillholes. (Svensk Kärnbränslehantering AB, Final Storage of Spent Nuclear Fuel, KBS-3 IV Safety (1983))

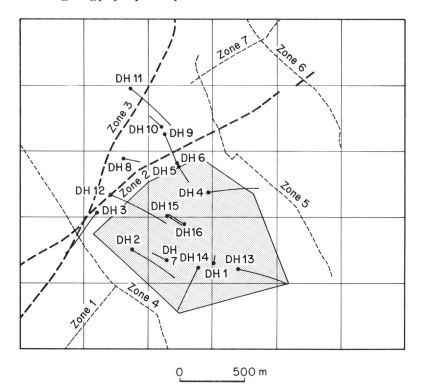

0   500 m

Figure 6.3 Map of fracture zones in the area of Figure 6.2 at a depth of 450 m, plotted from drillhole logs and examination of drillcores. Lines have the same meaning as in Figure 6.2. Shaded area is possible repository site. (Svensk Kärnbränslehantering AB, Final Storage of Spent Nuclear Fuel, KBS-3 IV Safety (1983))

drilling is the extent to which conditions at depths of a few hundred meters can be inferred from detailed mapping at the surface; the Swedish experience indicates that a fair correlation between the patterns of fractures at depth and on the surface is common but by no means universal.

Permeabilities at various depths were measured by packing off sections of the drillholes and injecting water from the surface. As was to be expected, the permeability of the rock itself is very low, often below the limit of measurement, while fractures and shear zones showed higher values. Since the measurements were made over depth intervals of several meters, the results were a composite of permeabilities for rock and fractures and showed much scatter. Overall however, in a majority of the areas examined, permeability became smaller with depth (Figure 6.4). This result is in accordance with experience elsewhere: a downward decrease of

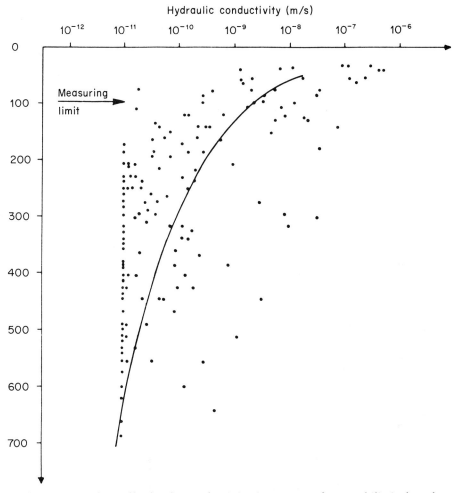

Figure 6.4 Values of hydraulic conductivity (a measure of permeability) plotted against depth for the drillholes shown in Figures 6.2 and 6.3. The dots in a vertical line on the left-hand side represent hydraulic conductivities below the limit of measurement, 10^{-11} m/sec. (Svensk Kärnbränslehantering AB, Final Storage of Spent Nuclear Fuel, KBS-3 IV Safety (1983))

permeability with depth is common in crystalline rock, but exceptions are frequent.

By detailed studies of this kind several of the areas examined were eliminated on the grounds of too much fracturing at depth or the lack of a volume of sufficient size to accommodate a repository and still leave adequate thicknesses of unjointed rock between the repository and major fracture zones. In four of the areas however, blocks of rock were found with only minor fracturing that would permit construction of a repository 1 km² in area protected by at least 100 m of sound rock on all sides. This was the minimum size felt necessary to contain the waste that will be generated over the next 40 years. Not one of the sites was ideal in all respects, and it is doubtful that any one of the four will be chosen as the site of the first Swedish repository. But the investigation showed that sites do exist in Swedish bedrock that are at least minimally satisfactory for repository construction, and this was the goal that the scientists had set themselves. There is ample time for further exploring, in the hope of finding a still better site, before a repository will actually be needed.

The work to date has shown no preference for one kind of crystalline rock over another. Some of the best sites are in granite, some in areas of gneiss and schist. A special effort was made to find a site in gabbro, with the thought that a mafic rock might be especially favorable because its ample content of ferrous minerals would ensure long maintenance of reducing conditions. No gabbro body was found however, that was both large enough and free enough from protesting landowners.

The Swedish studies of crystalline-rock disposal go well beyond the details of fracture geometry and measurements of permeability. The deep drillholes permitted collection of ground-water samples for analysis, which showed that water at repository levels had the desirable qualities of low total solutes, slight alkalinity, and low redox potential. The water varied somewhat from one kind of crystalline rock to another, but was generally noncorrosive. The ground water has been used in a variety of laboratory experiments designed to simulate expected conditions in a repository. Moreover, an underground laboratory has been set up in tunnels driven into granite near an abandoned iron mine (Stripa), where *in situ* experiments are in progress on many phases of repository construction and operation.

A final step in the exploration, when one or more specific sites have been selected as the most promising, will be the sinking of an exploratory shaft to or below the intended repository level. This will permit actual hands-on inspection of the rock to be excavated – the detailed mapping of joints and shear zones, measuring ground-water flow from the principal joints and testing of the feasibility of control by grouting, perhaps setting up *in situ* experiments to check the effects of this particular repository environment on canister metal and simulated waste. Such observations in a shaft or exploratory tunnel are essential, because no matter how carefully the surface work has been done, surprises are always to be expected underground. Mining practice has shown repeatedly that surface observations are inadequate for detailed predictions about conditions at depth.

Based on the field studies and experimental work, Swedish scientists have developed models to estimate the rate of escape of radionuclides to the biosphere under various assumptions (scenarios) about the timing and extent of canister failure. Even with the most extreme assumption of complete failure soon after the repository is closed (like the worst-case scenario of Chapter 4), the models indicate that the low solubility of the waste, coupled with sorption and dispersion as the ground water traverses the backfill and 100 m of sound rock, plus dilution from other sources of water near the ground surface, will be sufficient to keep doses even to maximally exposed individuals below regulatory standards.

The Swedish work is an impressive demonstration, apparently showing beyond reasonable doubt that sites suitable for repository construction can be found in crystalline rock — sites where the release of radionuclides can be adequately controlled by a combination of engineered and geologic barriers, and probably by either kind of barrier acting alone. The Swedes would be the first to grant that any such predictions involve uncertainties, that some details need confirmation. But on the basis of present knowledge it is hard to find serious flaws in the demonstration. To what extent the Swedish recipe for HLW disposal will be usable in other countries is an open question. Most countries would probably not choose to put so much copper underground, both because of the expense and because of the probability that it would appeal to a future generation as an easily accessible source of useful metal. Also the demonstration holds strictly only for a repository of small total area, far too small to accommodate the accumulating waste from reactors in countries with a large commitment to nuclear power. (In the United States for example, the minimum area for a repository site is estimated to be 8 km^2.) But certainly the Swedish scientists and engineers have given valuable guidance about the sort of exploration that will be needed wherever crystalline rock is being considered as a disposal medium.

Bedrock similar to that in Sweden covers large areas in eastern Canada and the northeastern United States. In Canada this rock is a prime target for possible repository siting, but in the United States no detailed study is planned until a second repository is needed (or later). Other countries where crystalline rock is regarded as a likely medium for repository siting include Austria, Czechoslovakia, France, India, Japan, and the United Kingdom.

6.3 BEDDED SALT IN TEXAS

Deep under the flat farmland of northwestern Texas are thick layers of salt, which constitute one of several salt accumulations being considered by the United States Department of Energy (DOE) as possible places for HLW disposal. Salt, as noted in Chapter 3, has long seemed a particularly attractive medium for repository construction because of its extreme impermeability, its ease of mining, its high heat conductivity, and its ability to flow plastically to fill openings. The United States has broad areas underlain by salt at appropriate depths, and the ongoing search for repository sites has revealed several places that appear to merit further study. At some of these places the salt has the form of domes – huge masses of salt punched upward into overlying sediments from a thick layer at great depth – while at others it retains the form of the sedimentary beds in which it was originally deposited. The northwestern Texas occurrence belongs to the latter group, but it will serve to illustrate some of the general considerations relevant to exploring for repository sites in any kind of salt.

The target bed in Texas is the Lower San Andrés formation, part of a thick sequence of evaporite beds formed by evaporation of an arm of the sea during the Permian period, roughly 250 million years ago. The salt beds are overlain and underlain by more normal marine sedimentary rocks, chiefly carbonates and shale, the whole sedimentary sequence forming a huge area of flat to gently dipping beds (the 'Permian Basin') that extends from southern Texas and New Mexico northward into Kansas. Minor undulations in the bedding outline subsidiary basins, some of which have been extraordinarily productive of gas and oil from strata beneath the salt and hence would be unsuitable for locating a repository. One basin in west Texas, the Palo Duro Basin, has produced very little despite extensive drilling, and it is in parts of this basin that exploration for repository sites has been concentrated.

On the surface the Palo Duro Basin is a featureless low plateau, devoid of outcrops except around its edges. The subsurface is known only from well-logging and examination of cores from the numerous holes drilled for petroleum exploration. Choosing a particular site for detailed study in such a region is difficult, because over a broad area the underlying sedimentary sequence has uniform characteristics and any one place has no obvious advantages over many others. To designate a tentative site, DOE scientists used such criteria as an appropriate depth to the salt bed, distance from the edges of the plateau, distance from population centers, and accessibility to roads and railroads. Many alternative sites will be available nearby, if further study shows that this one is not suitable.

There are no drillholes within several kilometers of the chosen site and the kinds and structures of rocks at repository depths must be guessed by interpolation on cross-sections constructed from the existing widely spaced holes. Fortunately the structure is simple enough (a very gentle dip south-eastward) and the strata are uniform enough that the sections can be very detailed. So when we learn that a 50 meter-thick salt bed in the Lower San Andrés lies at a depth of 700 m below the surface at the tentative site, we can be confident that the figures are accurate. With equal confidence we can describe the character of the bed, on the basis of cores from the nearest drillholes: the salt is mostly dark and impure, with abundant small patches of clay in and between the salt crystals, and with occasional continuous clay layers up to a centimeter or so thick; anhydrite occurs in patches and thick beds. This is not the ideal pure white crystalline salt that one visualizes in a salt mine (and that is found at some mines in the Permian Basin), but the bed has ample thickness for a repository and the amount of clay and anhydrite (roughly 3 and 7 volume %, respectively) is not large enough or continuous enough to seriously compromise the desirable properties of the salt.

If a repository were to be located here, a possible concern is erosion of the salt at the edges of the plateau. Such erosion is obviously going on today, and if it is rapid enough it might expose the repository before all of the contained radionuclides have decayed to low levels. Measurements of the rate of retreat of the plateau edges however, indicate that this possibility is remote. Another worry is the freshwater aquifer in Tertiary rocks above the Paleozoic section (the Ogallala aquifer): conceivably the aquifer could be contaminated, with disastrous effects on local agriculture, if engineering of the repository is careless; or Ogallala water might penetrate the repository and carry radionuclides down into the brine aquifer that is known to exist in strata under the salt beds. Use of standard engineering techniques should make it possible to prevent either of these calamities, but a demonstration is clearly desirable.

Still another cause for concern is local dissolution of salt in the immediate vicinity of the repository, a concern that arises whenever repository construction in bedded salt is contemplated. No evidence for appreciable present-day dissolution appears in available cores at the stratigraphic level proposed for the repository, but it is possible that mechanical disturbance during excavation, together with heat from the buried waste, might so alter the rock that ground water would gain access. The effect of heat is especially worrisome here because of the clay content of the salt: rising temperatures could dehydrate the clay minerals and so open passages along the clay and anhydrite seams which might permit entry of water. Pre-existing cracks in the salt might be opened by the heat and mechanical disturbance, and the cracks might not be healed by plastic flow of salt containing substantial amounts of impurities as fast as they would be in pure salt. Entry of water might also be promoted if the bedding in the repository area is broken by small faults that are not detectable on the surface or in the widely spaced drillholes; some inactive faults are known to exist in the general area, and fracturing on a smaller scale is not unlikely at any particular site.

One can argue that these possibilities are remote, and that even if water did collect in a repository it would probably do no harm because the surrounding impermeable salt would keep it from moving at an appreciable rate. But such deviations from ideal conditions have enough likelihood, especially in salt as impure as this, that they should be targeted for further study in the next stages of exploration – when deep holes have been drilled at the site itself, and later when the rock is accessible for detailed examination in a shaft.

Another aspect of the Texas site that seems favorable but needs confirmation is the apparently long path that ground water must follow from the site to reach the biosphere. If ground water does somehow invade the projected repository and dissolve some of the waste, and if it is not confined by the impermeable salt, its most probable route of travel is downward into a dolomite aquifer below the salt and then eastward along the bedding to the edge of the plateau 10–20 km away. From measurements of head and permeability in available drillholes, the time of travel from the repository to the edge of the plateau can be calculated as a few hundred thousand years, long enough to ensure adequate isolation of the waste. The validity of such a calculation can be questioned on the grounds that flow would be faster if it is channeled along major joints or fault zones. There is no evidence of such channeling, but the possibility will require further study of the hydrology.

Thus on the basis of present information a site in the Palo Duro Basin seems well suited for repository construction, but additional investigation is needed to satisfy possible objections. Among the bedded–salt sites that have been proposed, west Texas is somewhat unusual because of the impurity of the salt, but most of the geologic questions raised here are typical of this kind of repository environment. Typical also has been the response of the local citizenry: the mere suggestion that the Department of Energy considers this a particularly favorable place for detailed study has aroused a storm of protest.

Salt domes as sites for HLW disposal pose similar problems, except that ground-water movement along bedding planes need not be considered; the salt in a dome has been so thoroughly stirred by its long migration from deeper levels that it is effectively homogeneous. A question that does not arise in bedded salt is the possible effect of heat from a repository on the mechanical stability of a dome: could heating so increase the plastic mobility of the salt that the entire dome would rise further into its surroundings, ultimately bringing the repository to the surface? Like many of the questions about bedded salt, this one seems to address a remote possibility, but one that nevertheless needs serious consideration.

In other countries fortunate enough to have large deposits of rock salt, this medium figures prominently in plans for HLW disposal. Especially notable are developments in the Federal Republic of Germany, where a shaft for an intended repository is being sunk into a salt deposit at Gorleben, near the border of East Germany. The salt here takes the form of a 'salt anticline', in effect an elongated incipient salt dome – a bed that has been thickened by plastic rise of the salt due to the weight of adjacent overlying rocks, forming a wide salt ridge with superjacent beds arched over it. Like the salt in west Texas, the north German salt beds are Permian; the same salt strata underlie parts of Denmark, Poland, and the Netherlands and, in these countries also, they are being considered for waste disposal. Beds of older salt (Devonian) in some of the western provinces of Canada would have possibilities as disposal sites, but the Canadians have so far given more attention to the crystalline rocks of the Canadian Shield.

6.4 TUFF IN SOUTHERN NEVADA

The southern part of Nevada has much appeal for repository siting because its combination of prevailing high temperatures, low rainfall, and substantial relief means that the permanent water table is very low, more than 500 m below the surface in much of the area. The opportunity is thus presented to locate a repository in the unsaturated (vadose) zone well above the water table. In such a position the repository would be free of the possibility of continuous flow of ground water through it and the consequent steady dissolution of radioactive material. One of the major worries that beset plans for repository construction in most areas would be eliminated.

This region also boasts a peculiarly attractive kind of rock for repository siting: consolidated volcanic ash or tuff. During a period roughly 40 to 10 million years ago a succession of rhyolitic volcanoes erupted huge ash clouds that formed layers of ash totaling more than 2000 m in thickness. Most of the layers are ash-flow deposits, consisting of material laid down by turbulent flows of hot gas and glassy particles, so hot when it accumulated that particles in the interior of a layer were partly fused together to form 'welded tuff'. The glass was unstable, and under the influence of heat and long standing it has partly devitrified to form a very fine-grained crystalline aggregate of feldspar and cristobalite. Interbedded with the ash-flow tuffs are layers of ash-fall material, ash that accumulated more slowly by simply falling through the air and thus was cooled before deposition. As a repository medium, welded tuff has the advantages of ready minability by standard methods, a capacity to maintain openings indefinitely with little or no support, and immunity from the effects of moderate heating. The unwelded and partly welded tuff nearby has the further desirable property of strongly sorbing any radionuclides that might escape; not only the ash particles themselves, but even more the clays and zeolites that are common alteration products of the tuff, are effective retardants for most ions.

The geology of southern Nevada is complicated by numerous vertical or near-vertical faults and by the discontinuous, lenticular form of most of the tuff deposits. Choosing a good site requires finding a tuff bed with the appropriate combination of welded and nonwelded material, then locating a part of the bed that has the necessary thickness, is well above the water table, and is cut by few faults. The site tentatively selected by the Department of Energy for further study is in an ash-flow deposit (Topopah Springs member of the Paintbrush tuff) about 350 m thick underlying a desert range (Yucca Mountain) at the edge of the Nevada Test Site. This location has the advantage that parts of the near-surface material at the Test Site are already contaminated by radioactive debris released by underground testing of nuclear weapons, so that minor additions from a remotely possible malfunctioning of a repository would be less serious here than at most sites elsewhere. It is also situated far from population centers, but close enough to major roads and rail lines that development of the necessary transportation system would not be difficult. With all these favorable attributes, it is small wonder that southern Nevada stands high on the list of areas that the DOE considers worthy of increased attention.

On the other side of the ledger one can point to some geologic questions that are at least cause for concern. First is the matter of volcanism: this region has obviously experienced intense eruptive activity in the fairly recent geologic past; how can we be sure that activity will not be renewed during the next few thousand years? Isn't an explosive eruption like those that formed the tuff beds a real possibility? Even without an eruption nearby, isn't there danger that the hot fluids often found in volcanic areas could generate radioactive hot springs? The best answer is simply the well checked radiometric dating of rocks in the vicinity: the most recent of the rhyolitic eruptions was more than 8 million years ago, and the last eruption of basalt (8 km from Yucca Mountain) at least 20 000 years ago, so that the chance of recurrence in the near future seems slight.

More disturbing is the possibility of major tectonic movement, since the area is cut by many faults that were active during the Tertiary. Evidence for recent seismic activity is meager: a few faults are marked by small subdued scarps suggesting Pleistocene or Holocene displacement, and minor earthquakes (Richter magnitude less than 4) have been recorded in the general vicinity. Some tectonic disturbance is certainly to be expected in the future, but probably not of sufficient intensity to damage a repository; and surely in locating a repository any fault with indications of recent movement would be avoided. Volcanism and seismicity must be reckoned as disadvantages of southern Nevada, but the chance of either causing major harm to a repository is small.

Although placing a repository in the unsaturated zone a couple of hundred meters above the water-table would seem to make it immune to serious invasion by ground water, this is not beyond question. Rain falling on the southern Nevada desert generally penetrates only a few centimeters into the ground, then most of it returns to the surface by capillarity and evaporates. A little, especially during the rare major storms, percolates downwards through the unsaturated zone to the water table (although much of the water-table recharge comes from rainfall in higher areas to the north). Scanty evidence suggests an average flux of water through the vadose zone of a few tenths of a millimeter per year. Thus a little water will doubtless occasionally trickle through a repository, and over many thousands of years might be sufficient to corrode canisters and dissolve part of the waste. Any dissolved material, on reaching the water table, would move horizontally through aquifers in the tuff to seepages in desert basins a few tens of kilometers away. The time of travel, spent mostly in the downward motion through 200 meters of tuff in the unsaturated zone, has been variously estimated as between 9 000 and 80 000 years. One can take comfort in thinking that dissolution of waste will be delayed for a long time and that the amount of water will be small; on the other hand, the water will almost certainly be oxidizing, since it comes directly from the surface, and hence will be abnormally corrosive. Thus a scenario can be imagined by which, even in this most favored sort of repository environment, radionuclides might at some time reach the biosphere in hazardous amounts. The scenario is not very plausible, but its plausibility needs testing by further work on the general hydrology at the site and on details of the movement of water in the vadose zone.

The Yucca Mountain site is manifestly attractive, but its suitability will need checking, as will the suitability of sites in Texas and in Sweden, by additional study aided by deep drilling and ultimately by the sinking of an exploratory shaft.

6.5 SUMMARY

The three kinds of sites we have examined – in crystalline rock, in salt, and in tuff – illustrate the geological considerations that arise in any effort to find a specific place for a waste repository. The first two are examples of kinds of sites that can be found in many countries. The third may be unique, but it highlights the sorts of questions that need answering about any site where placement of a repository above the water-table is possible, and also about any site to be considered in an area of geologically recent tectonism and volcanism.

As the three examples show, the search for repository sites involves first a reconnaissance of the geology over a broad region, then the choosing of smaller areas for detailed mapping, then narrowing the search to still smaller areas where subsurface information is sought from drillholes, either existing ones or holes drilled for this special purpose. Accompanying the geologic study must be investigation of the hydrology, first on a regional scale and then in detail with data obtained from deep drillholes at the possible repository sites. Geophysical measurements from the surface and from the drillholes provide more details about underground rock structures and ground-water movement. Even with the best surface mapping and the most complete drillhole logging however, some geologic uncertainties will remain.

To resolve the uncertainties, the final step of the exploration must be the sinking of a shaft to repository depths and the opening of exploratory tunnels, so that rock structures and water movement can be studied in place. At this step one must be prepared for surprises, because long experience in mining shows that the most meticulous study of the ground surface can give only an approximate idea of what may be found at depth. If the surprises are minor, construction of a repository may be authorized and the shaft can be incorporated into its design. If subsurface conditions show major differences from those predicted, it may be necessary, even at this late stage, to abandon the site and seek another.

7

Natural analogs

7.1 USES OF ANALOGS

A repository for high-level waste, we have emphasized repeatedly, will be a geologic anomaly: a huge mass of material containing unusual elements, enclosed in rock some hundreds of meters underground, heating its surroundings for hundreds of years and emitting intense radiation for many thousands. Nothing of the sort exists naturally on the earth, and nothing like it has been constructed previously, so we have no good example with which to compare predictions about its behavior. Nevertheless, natural environments can be found that resemble some aspects of a repository – places where geologic inference tells us that materials somewhat similar to those in a repository have existed in contact with ground water at slightly elevated temperatures for thousands or millions of years. Such places supply, in a rough way, the ingredient that is missing from our models: visible evidence about the behavior of repository materials over very long periods of time. The evidence is necessarily equivocal, because natural conditions can never duplicate precisely the expected environment of a repository. Sometimes the evidence permits us to make semiquantitative estimates of rates; more commonly it is only qualitative. But since predictions about repository performance are based on outrageous extrapolation from short-term observations and laboratory experiments, any help we can get from evidence of this sort is welcome as a test of our reasoning.

We cannot hope to find a natural situation analogous to an entire repository, or even to a major part of one. At best we will be seeking, either in natural rocks or among archeological remains, materials like some of the individual substances that make up a repository which have existed for long periods under conditions similar to those we postulate in the repository environment. Finding such materials and judging what inferences can safely be made from them requires a good deal of geologic ingenuity. It is all too easy on the one hand to overlook pertinent evidence, and on the other hand to be overly eager in drawing unwarranted conclusions. Out of the multitude of natural analogs that have been suggested, we choose a few examples for brief discussion.

7.2   ANALOGS FOR CANISTER MATERIALS

The metals currently most favored for canister construction are iron, copper, and titanium. Because titanium is an active metal that has only recently been produced commercially in any quantity, a search for analogs either in rocks or ancient tools would be pointless. Copper and iron however, were widely used by primitive societies and in the ancient civilizations, so that many samples which have been buried for a few thousand years are available for study. Native copper is also known in many ore deposits, but natural occurrences of native iron are too rare to be of much help.

The dating of buried copper tools and measurements of the amount of corrosion permit rough estimates of the time a copper canister can be expected to survive. The estimates are widely variable of course, because of differences in conditions of burial, but they confirm predictions made from Swedish experimental work that copper canisters with walls 10 cm thick would not be breached in several hundred thousand years and very probably not for a million. Still longer survival times for copper canisters are suggested by studies of the famous native copper deposits in the Precambrian of northern Michigan. Part of the evidence here is the simple fact that metal remains after exposure to ground water for much of the time since the late Precambrian; more quantitative corrosion rates can be derived from study of the slightly corroded surfaces of copper boulders that were detached from outcrops by Pleistocene glaciers and scraped clean at least 8000 years ago.

The iron alloys used in ancient times were different from those commonly suggested for canisters, but the cast iron planned for use in Switzerland (and possibly in the United States) is sufficiently similar to make corrosion rates of artifacts relevant. The preserved iron objects show more alteration than copper from similar environments of course, but the estimated rates suggest that corrosion would not breach the planned thick-walled iron canisters for at least a thousand years. Thus natural analogs of two proposed canister metals provide striking confirmation of the survivability that has been predicted from theory and from laboratory experiments.

7.3   ANALOGS FOR BENTONITE IN BACKFILL

The composition of the material to be used as backfill around canisters and in repository tunnels is still a subject of debate, and will doubtless vary from one repository to another. The crushed rock removed during repository excavation will probably be a major constituent, but plans commonly call for the use of clay as an additive or, in some repository designs, as the bulk of the backfill material. The clay generally favored is bentonite, a smectite-rich clay derived by alteration of volcanic ash. The desirable attributes of bentonite are its ability to swell on contact with water, so that it fills cracks and open spaces tightly, and its strong sorptive capacity for cations. Whether it will retain these properties in a repository during the long times required for HLW isolation is questionable. Observations of the behavior of bentonite under natural conditions help to reduce the uncertainty.

The existence of bentonite in beds as old as mid-Paleozoic is ample evidence for its stability at low temperatures, but both experiments and observations in nature show that it gradually changes to illite at temperatures above $100°$ C if a source of K^+ is present. The change to illite may not be seriously damaging to a repository, but it would certainly make the clay somewhat more permeable, less expansive and less sorptive. A demonstration that the change would not progress far in the time needed for waste isolation would increase confidence in the value of bentonite as a constituent of the repository system. Finding a place where the demonstration seems entirely convincing is not easy, because values of the two variables (temperature and concentration of K^+) in the geologic past can only be approximated, but study of deeply buried smectite-rich sedimentary rocks in the south-central United States has shown that the transformation to illite is very sluggish provided temperatures do not rise far above $100°$ C. For repositories like those planned in Sweden and Switzerland, where expected temperatures will be kept below $100°$ C, the use of bentonite seems fully justified. If maximum temperatures will rise above $200°$ C, as is expected in current planning for repositories in the United States, the long-term stability of bentonite is less certain.

7.4   ANALOGS FOR THE WASTE FORM

Spent fuel consists essentially of uranium dioxide pellets, and the obvious analog is a deposit of the same compound in nature – the common uranium ore mineral uraninite, or the less pure variety pitchblende. The spent fuel of course, contains elements not present in the natural mineral: the transuranic actinide elements, which probably substitute for uranium in the UO_2 crystal structure, and fission products, which are partly within the crystals and partly on crystal margins. The spent fuel differs from the natural material also in that its crystal structure may be partly disorganized by the intense radiation field, and hence may be somewhat more soluble. Despite the differences, study of uraninite ore and its surroundings should provide clues as to the behavior of spent fuel in a repository, at least with regard to the mobility of uranium itself.

A number of such studies, particularly at the famous ore deposits in Saskatchewan and northern Australia, have shown that uranium behaves generally in accord with predictions that can be made from its chemistry. As long as conditions remain reducing, the element is practically immobile; measured concentrations in ground water are in the neighborhood of 1 ppb (part per billion, or $g/10^9$ g), similar to UO_2 solubilities measured in the laboratory and well below the maximum concentration permitted in water for ordinary use. Where there is evidence of migration in ground water into the rock surrounding an ore accumulation, the movement of uranium is restricted to distances of several meters, even in the two or three thousand million years since early in the Precambrian.

If oxidizing ground water has been in contact with the ore, the situation is entirely different. Uranium is readily oxidized to the hexavalent state, and its mobility in this form is often shown dramatically by the brightly colored uranyl minerals in outcrops above drab, dark-colored primary ore. From solutions of hexavalent uranium the oxide may be reprecipitated if the solutions move from an oxidizing into a reducing environment. This sequence is often noted in sandstone ores: where oxygen-rich water from the surface moves into a bed containing disseminated uraninite plus reducing materials (sulfides and/or organic matter), the dissolved uranium travels along the bed until oxygen in the water is exhausted, then reprecipitates; the point of reprecipitation is marked by a conspicuous curved band ('roll front') with yellow and orange colors on the upstream (oxidizing) side and black oxide on the downstream side. One can imagine a similar structure developing in a repository: if ground water is in contact with spent fuel, radiolysis might well create local oxidizing conditions (Chapter 5) and some uranium would dissolve; but movement of the ground water would carry the uranium into backfill or fresh rock where reducing material (sulfides, ferrous silicates) would change it again to the insoluble oxide. Such 'oxidizing fronts' are often postulated in scenarios of radionuclide release.

Thus observations of uranium behavior in and around ore deposits gives added assurance to the conclusions from theory and laboratory experiments that long-distance migration of uranium away from spent fuel in a repository will be limited to amounts well within the allowable concentrations set by regulatory agencies. Because the chemistry of plutonium and neptunium is similar to that of uranium, especially in that both elements also form dioxides that are extremely insoluble under mildly reducing conditions, it is reasonable to suppose that these elements too would escape from spent fuel in only minimal amounts. This conclusion is of course less certain than the conclusion for uranium.

For the other planned waste form, borosilicate glass, finding a natural analog is more difficult. One thinks immediately of natural glasses like obsidian, or glass objects preserved from ancient times, but the compositions of these glasses are so different that their usefulness as analogs is questionable. Basalt glass is most similar in silica content, and hence is probably the closest analog. The slowness with which natural glasses devitrify suggests that devitrification of borosilicate glass in the time needed for waste isolation is not a serious concern; and even if devitrification occurs, the slight change in solubility produced by the observed devitrification in natural glasses indicates that this is a matter of minor consequence. Hydration of natural glasses on prolonged contact with ground water however, suggests that hydration of glass in a repository could eventually alter its structure and increase its solubility, but the seriousness of this change is hard to predict. Models of repository behavior often assume that radionuclides in glass are readily accessible to ground water, and that their mobility is a function chiefly of solubility and sorption rather than inertness of the waste form; since such models commonly show that most nuclides would not reach the biosphere in unacceptable amounts, the lack of a satisfactory analog for borosilicate glass is probably not very serious.

7.5 ANALOGS FOR RADIONUCLIDE MOVEMENT

The travel of radionuclides through rock away from a repository, as conditioned by the solubility of their compounds and their sorption characteristics, can be simulated by the observed movement of more ordinary materials. For some of the nuclides in HLW, the obvious analogs are simply nonradioactive isotopes of the same element; for example, the behavior of ^{90}Sr and ^{137}Cs should be similar to that of ordinary strontium and cesium. The expected trapping of these two radioisotopes in most rocks by sorption and ion exchange is amply confirmed by observation of the nonradioactive analogs, and the expected non-trapping of ^{129}I is attested by the observed mobility of iodine in ordinary rocks. For technetium and the transuranic elements however, no nonradioactive isotopes exist, and the amounts in nature of the radioisotopes themselves are so exceedingly minute that they are of no help as analogs of the elements in a repository. For these elements then, we must have recourse to analogs among elements that are chemically similar.

With americium and curium this is simple, because only a single oxidation state is possible for these elements in repository environments, and the chemistry of Am(III) and Cm(III) is similar to that of the light rare-earth elements (La–Sm). Neodymium in particular, with an almost identical ionic radius and similar hydrolysis constants, is a near duplicate of the two actinides; the immobility of this element in ground water near rare-earth deposits is good evidence that both americium and curium will be trapped before moving far from a waste repository.

For elements with multiple oxidation states, finding suitable analogs is more troublesome. As noted above, uranium in some respects is a good stand-in for neptunium and plutonium, but its chemistry is sufficiently different that other analogs seem more informative. For the trivalent state of these elements, just as for Am(III), much of the chemical behavior is duplicated by the rare-earth elements. In the tetravalent state thorium is a fairly satisfactory analog, because it too forms a very insoluble dioxide and has similar formation constants for its complex ions; thorium differs from the heavier actinide elements in that its only oxidation state is $+4$. The analogy with thorium is better for plutonium than for neptunium, because NpO_2 is more easily oxidized than PuO_2 and Np forms more stable complexes. The higher oxidation states of these elements are best likened to those of uranium, but the analogy is far from perfect. For technetium a possible analog is rhenium, which shows a similar alternation between oxidation states $+4$ and $+7$ with changes in redox conditions, but this analogy has not been well studied.

An example of the sort of conclusions that can be drawn from natural analogs about the mobility of the transuranic elements is provided by a deposit of thorium and rare earths in southern Brazil. The deposit is located near the top of a small hill (Morro do Ferro) on a plateau underlain chiefly by alkalic igneous rocks; the thorium and rare-earth elements are in large part disseminated through a deep clay soil formed by tropical weathering of the igneous material. Ground water that seeps through the soil drains into a small creek at the base of the hill. If the ore elements can be mobilized by ground water and surface water, this should be the ideal place to find them in the drainage water, yet the observed concentrations of all of them are very low. Conditions here are obviously different from those expected in a repository, because all materials are highly oxidized; but since both thorium and the rare-earth elements have only a single oxidation state ($+4$ for Th and $+3$ for the rare earths, with minor exceptions for a few of the latter), their behavior should be the same in the reducing environment of a repository. Transuranic elements in a repository then, should show similar behavior, provided their oxidation states are the same; americium and curium will certainly be in their trivalent states, plutonium almost as certainly tetravalent, and neptunium probably, but somewhat less certainly, also tetravalent. The failure of the analog elements to move in appreciable amounts under oxidizing conditions is good evidence that the corresponding radioisotopes will be similarly restricted in a reducing environment underground.

A caution about such reasoning should be emphasized. All of these elements form many stable complexes, both with inorganic ions and with organic compounds in solution. Stability constants are known for many of the complexes. The constants for radioisotopes and corresponding analog elements are mostly similar, but a few are notably different. So it is possible, although not very likely, that formation of some of the lesser known complexes might make the mobility of the radionuclides greater than the behavior of the analogs would imply.

7.6   OKLO

One occurrence of uranium ore that should seemingly have special significance as a natural analog is a deposit at Oklo, in the West African country of Gabon. The ore consists of uraninite and pitchblende in pockets of shale in a Precambrian sandstone. Routine analyses in the early 1970s by the French operators of the Oklo mine gave strange results: the isotope ratios $^{235}U/^{238}U$ in ore from some parts of the mine were much lower than normal. This was odd, because no ordinary natural process can produce appreciable separation of the two isotopes. French scientists who were called on for an explanation suggested that ^{235}U might have been consumed in a natural fission reaction, and intensive study of the ore showed that the expected products of a nuclear reaction were indeed present. Oklo then, has the remains of a natural nuclear reactor, a reactor that operated intermittently, at a depth of about 3500 meters, for more than half a million years at a time about two billion years ago. Existence of the reactor required special circumstances: uranium ore of the proper grade and distribution in the sediment, absence of elements that would absorb neutrons, and water present part of the time in just the right amounts to serve as moderator. Such a combination of conditions would be very rare, and Oklo is probably unique. Extensive research in Precambrian rocks elsewhere has failed to locate any sign of a similar fossil reactor.

At several places in the Oklo mine, all the products of a nuclear reaction were generated underground and then remained in a ground-water environment for some two billion years. The original radionuclides of course, have long since disappeared, but the stable isotopes resulting from their decay are readily detectable. Offhand this deposit would seem to be an ideal analog for a waste repository, a place where predictions about rates of movement of various radionuclides could be given a rigorous test.

Unhappily the analogy turns out to be far from perfect. The uraninite ore has indeed survived with little evidence of dissolution over the very long time since the reaction stopped, as would be expected from theory and experiment and from many observations like those noted above at other deposits of uranium ore. But evidence about migration of the reaction products is equivocal. The heavy actinide elements were retained, as would be expected, but the lack of movement may well be due more to retention in the uraninite crystal structure than to insolubility and sorption from moving ground water; uraninite crystals are better preserved here than those in spent fuel rods because of a less complete nuclear reaction, so that reaction products would be less accessible to ground-water attack. More mobile radionuclides (^{90}Sr, ^{137}Cs, ^{99}Tc, ^{210}Pb) moved away from the reaction zones, by relative amounts roughly in accordance with expectation, but not enough is known about the changing conditions of Eh, pH, and temperature over two billion years to make possible a quantitative test of agreement with prediction. It is disappointing that this one place on earth where we actually see in a natural setting the results of long exposure of reaction products to ground water can give us so little precise information.

7.7   DISCUSSION

Study of natural analogs helps to corroborate some of the inferences about repository performance that we drew earlier from theoretical considerations and short-term experiments. Support from analogs is largely qualitative; only rarely do we find it possible to state even semi-quantitative conclusions, and these are liberally hedged with qualifications. Nevertheless the information that analogs supply about long-term processes, vague though it often is, provides a useful supplement to the estimates we make from observations and experiments that are necessarily limited by the lack of time available for laboratory work.

Reasoning from analogy involves obvious assumptions, as does reasoning from the theoretical background and from laboratory experiments. In effect, we try each of the three approaches, and find that in each some conclusions are not as firmly based as we could wish. It will probably never be possible to state with finality the amount of radioactive material that will be released from a repository at a given time in the future, but we come closer to this goal by using a multitude of approaches. Our previous conclusion that escape of radionuclides in unacceptable amounts from a well-constructed repository is highly unlikely, surely gains plausibility from our findings that an approach through analogs leads to the same sort of expectations as the other approaches we have tried.

__8__

The subsea-bed option

8.1 THE OCEAN AS A DUMPING-GROUND

Among the alternatives to mined geologic disposal listed in Chapter 3, the most attractive is burial of HLW under the floor of the deep ocean. The immediately obvious merit of this suggestion is removal of the waste far from any inhabited area; one immediately obvious disadvantage is the need for complex political negotiations to legitimize such use of international waters. Despite this potential difficulty, subsea-bed disposal is appealing enough to have been the subject of serious study in several nations for more than a decade; in the United States it is widely regarded as a possible substitute for additional repositories on land a few decades hence. In this chapter we return to the subsea-bed option for a closer look.

Dumping in the ocean has always seemed a good way to get rid of unwanted rubbish. The volume of the ocean is so huge, the volume of human generated waste so relatively small, that seemingly one could depend on simple dilution to quickly reduce the concentration of any sort of toxic material in the waste to undetectable levels. This argument might be valid if mixing in the ocean were sufficiently rapid and widespread, but such rapid mixing cannot be assured. Several unhappy incidents have shown that toxic material added to coastal waters can remain undispersed long enough to be ingested by micro-organisms and then concentrated as it is passed along up the food chain to higher organisms. It is generally not the concentration in sea water itself that poses a hazard – this can be kept low by proper spacing of the dumping – but the step by step increase in concentration of toxic compounds in bodily tissues by metabolic processes as one kind of organism consumes another, up to the fishes and shellfish that are often a major part of the human diet. If radioactive isotopes are present in the waste, even minute traces can be concentrated in this way to become a menace to coastal communities.

It is universally agreed therefore, that radioactive waste of any kind must be kept out of near-shore waters, and that high-level waste must not be placed anywhere in the open ocean. Disposing of low-level waste in deeper water offshore – particularly waste containing only isotopes of short half-life that decay to harmless levels within a few decades – is a practice that was followed by the United States and several European countries for many years with no evident ill effects. The United States discontinued such ocean dumping of low-level waste in 1970, and European countries in 1982, but resumption of the practice is being seriously considered. For coastal nations that lack good sites for low-level waste on land, this method of disposal seems a reasonable alternative – provided that the dumping is far from shore, and that no high-level material is included.

The proposal to put HLW under the floor of the deep ocean is an entirely different disposal concept. The waste would not be simply tossed overboard into the sea, but in the form of an insoluble solid would be enclosed in metal sheaths, much like the canisters proposed for disposal on land, and buried 30 or 40 meters under the surface of the fine-grained sediment that covers much of the deep ocean bottom. For any radionuclides to reach the human environment would then require that the metal containers corrode, the waste form dissolve, the radionuclides move through the highly sorbent sediment to the sea floor, and then be transported by currents for hundreds of kilometers to shallow water. These barriers would presumably so slow the movement, and so disperse the nuclides, that they would not reach coastal waters in any quantity for the hundreds of thousands or millions of years needed for decay to render them innocuous.

8.2   FINDING A DISPOSAL SITE IN THE OCEAN

In the vast expanse of ocean bottom available for this kind of disposal, how does one set about locating an appropriate site? The problem seems more difficult than finding a place for disposal on land, simply because outcrops of rock and sediment are not readily accessible for study. A possible choice that comes to mind immediately – and one that is often suggested by laymen – is an oceanic trench, one of the deep troughs that are commonly present along subduction zones. If HLW is buried in the loose sediment that accumulates in such a trench, subduction would carry it down into the earth's mantle and thus remove it not only from the biosphere but from the entire crust. Surely no more permanent kind of disposal can be imagined.

The difficulty with this suggestion, as noted in Chapter 3, is our uncertainty about the mechanism of subduction at any particular place. While it seems clear that some sediment is generally carried downward with a subducting plate, abundant evidence shows that along some subduction zones a part of the sediment has been pushed against the adjacent continent or island arc instead. Radioactive material buried in the sediment could thus become exposed on land or in shallow water, and might be a long-term hazard to future coastal inhabitants. This possibility is sufficient to rule out subduction zones as likely disposal sites. We can also eliminate from consideration parts of the ocean near mid-ocean ridges and active volcanoes, simply to avoid uneven topography and possible complications from tectonic or magmatic activity.

More favorable sites can be found in the interiors of the great oceanic plates, far from the edges where plate material is being formed, destroyed, or in motion against other plates. Such plate interiors are probably the most tectonically stable environments to be found anywhere on the earth's surface. The plates are moving at rates of a few centimeters per year, so that a site located several hundred kilometers or more from a plate margin would not be carried near the edge for at least a few million years.

Then one would consider the kinds of sediment on different parts of the deep-sea floor in which the waste might be buried. The sediments are wide ranging in composition: some are made up largely of the calcareous shells of tiny organisms, some of siliceous shells, some of inorganic materials; some have abundant organic matter, some only traces; some are characterized by abundant nodules of iron and manganese oxides. For waste-disposal purposes the most suitable sediment is the clay, particularly the iron oxide-rich variety called 'red clay', that covers large expanses in the deeper parts of the sea (depths below 5 km): this is inorganic material formed largely from dust blown far out to sea from the continents, which accumulates at rates estimated between 0.1 and 10 mm per thousand years. Two prominent constituents are clay minerals and ferric oxide, substances that would be particularly active sorbents for many of the radionuclides that might escape from buried waste. The low content of organic matter (generally less than 0.5%) would minimize concern about mobilization of radionuclides by formation of organic complexes, and absence of manganese nodules would eliminate possible future disturbance by attempted commercial utilization of the sediment. A further advantage of this sediment is its remarkable uniformity: many cores obtained from the deep ocean have shown that characteristics of the clay are similar from place to place and have not changed appreciably over many millions of years, despite the radical changes in climate during the Pleistocene. Thus the nature of the sediment over a large area can be inferred from a few drillcores, and can be confidently predicted to remain the same through changes of climate in the long term.

Basing their choice on considerations of ocean depth, nature and thickness of sediment, ocean-bottom topography, and distance from plate margins, oceanographers in recent years have designated several areas in the Atlantic and northern Pacific oceans that seem especially suitable for HLW disposal. Continuing research on the possibilities of developing a subsea-bed disposal operation will be concentrated in these areas.

8.3 TECHNIQUES OF SUBSEA-BED DISPOSAL

The best procedure for burying HLW in the deep-sea sediment is still under discussion. Certainly the waste (in the form of either spent fuel or glass from reprocessing waste) will be enclosed in metal containers, transported by rail or truck to an ocean port, and carried by ship to the disposal area. Several alternative ways of getting the containers into the sediment have been suggested (Figure 8.1). A method attractive for its simplicity is to put waste canisters into sheaths shaped like a projectile, or to give the canisters themselves the form of a projectile, and drop them overboard. Calculations and experiments indicate that free fall through five or six thousand meters of sea water would give the projectiles enough speed to bury themselves the

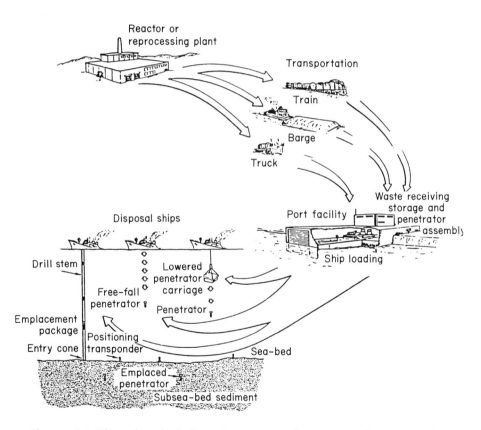

Figure 8.1 The subsea-bed disposal concept. Three ways of inserting the projectile-shaped canister ('penetrator') into the sediment are shown: free fall from the ship, fall from a lowered carriage to provide better control of speed and position, and emplacement in a drilled hole. (Office of Technology Assessment, U.S. Congress, Staff Paper (1985))

required 30 or 40 meters into the loosely consolidated, brine-saturated sediment. For further assurance that sufficient depth will be reached, explosive devices could be used to increase the speed of entry into the sediment. Still greater certainty that the waste containers would be properly emplaced could be provided by more elaborate ocean-engineering techniques: holes could be drilled into the ocean bottom materials and the canisters could be inserted mechanically, just as they would be in the tunnels of a mined geologic repository on land. With this last method it would even be possible to retrieve the waste at a later date, if this should ever be necessary, by relocating the drilled holes and extracting the containers. The different techniques of getting HLW into the sediment differ widely in cost, the last being by far the most expensive; a choice among them would depend, presumably, in part on the expense and in part on the importance attached to having the waste containers at a particular depth and at known locations in the clay.

A buried container, surrounded by the interstitial sea water of the sediment, is in a highly corrosive environment. The salt water in contact with its surface will be hot, the specific temperature depending on the nature, age, and amount of waste; according to one recent proposal these variables would be controlled so that the temperature will rise to a maximum of about 250° C within three years of emplacement, and then decline to less than 100° C within 40 years. Conditions will be mildly oxidizing, because bottom water at the ocean floor in most places contains a little dissolved oxygen. Finding a metal that will stand up to such conditions is not easy, but recent work indicates that an alloy of titanium (Ticode 12) can provide the required resistance for at least a few centuries.

Thus, there seems little reason to doubt that presently known techniques are adequate to accomplish subsea-bed burial of HLW, if a decision is reached that this would be an appropriate disposal method.

8.4 PROBLEMS OF SUBSEA-BED DISPOSAL

What difficulties can be anticipated in getting HLW safely under the sea floor, and how effectively would the waste be isolated?

Transportation would surely pose more problems for the subsea-bed alternative than for a disposal in a mined repository, since the waste would have to be moved long distances overland, then transferred to an ocean-going vessel and carried hundreds or thousands of kilometers to the disposal area (Figure 8.1). The greater amount of handling obviously means greater hazard to operating personnel, and the longer distances of travel mean more chance of accidents en route. An objection often urged against subsea-bed disposal is the statistically greater probability of accidents at sea than on land. Perhaps the consequences of a marine accident would be less severe, since even in the worst imaginable accident a shipload of waste would be simply lost in the ocean rather than spread over a terrestrial landscape, but the objection is still a valid count against this method of disposal.

Metal containers buried under the sea floor would be in a more hostile environment than that expected in most of the planned terrestrial repositories, and eventual breaching by corrosion is inevitable. When the waste is in contact with warm salt water, some radionuclides will surely dissolve. How fast can they be expected to move away from the waste? This is hard to predict, because our knowledge of conditions within deep-sea sediments is limited. High temperatures near the waste might be expected to cause convection in the interstitial fluid, which could carry dissolved material up to the sediment–sea water interface; but calculations show that convectional motion after the time when canisters are breached would not be significant. The principal way for radionuclides to move is by diffusion. The expectable rate of diffusion, according to laboratory experiments simulating subsea-bed conditions, would be sufficient to carry ions from exposed waste at a depth of 30 m to the sea water interface in roughly 1000 years. Radionuclides that form cations however, would be greatly retarded by sorption on clay minerals and iron oxides, so that only anions like I^- and TcO_4^- would escape in any quantity. In the bottom water they would be picked up by slow-moving currents and might be carried long distances, conceivably even into the shallow water of coastal regions. Or the nuclides could be consumed by bottom-dwelling organisms, and might eventually move up the food chain into creatures that would be eaten by humans. Attempts to model such unlikely scenarios, admittedly based on scanty data, indicate that radiation doses to individuals on land would be orders of magnitude smaller than doses from natural radiation.

Even more difficult to handle than the technical uncertainties are the geopolitical problems that would arise in establishing a subsea-bed disposal site. It seems likely that nations with a well developed nuclear power capacity would be able to reach some sort of agreement, since all will have similar problems of disposal and all could use the same general area if this method proves satisfactory, but disputes about territory and about disposal practices are readily imagined. More troublesome would probably be the countries without a nuclear power industry, which for one reason or another tend to exaggerate the risk of contamination near their shores from a waste-disposal site thousands of kilometers distant. But no international conference on the subject has ever been held, and national viewpoints can only be guessed.

Subsea-bed disposal has the obvious attractions of getting the waste far from land, of assured geologic stability, and of an environment that seems admirably suited to trap escaping radionuclides by sorption. Aside from political problems, the most serious of the known disadvantages are the greater danger of accidents and the greater risk to operating personnel in getting the waste from the reactor to the disposal site. There are many uncertainties however, simply because we know less about this environment than about sites on land. It is generally agreed that we need better information about the long-term effect of heat on red clay, about the long-term behavior of metals and waste forms in contact with hot sea water, about the rate at which sediment would close in to fill the hole above a buried projectile, and about the ability of red clay to sorb the important radionuclides. The information we have on these topics is all favorable, and one can say with some confidence that burying HLW in well-chosen sites on the deep-ocean floor gives every promise of isolating the radionuclides satisfactorily for the necessary long future times. In the opinion of some experts, the subsea-bed option is preferable to any of the possibilities on land. But until the uncertainties have been reduced by further research, the safest present course would seem to be staying with plans for disposal in land-based repositories, simply because here our knowledge base is so much more extensive.

8.5 HIGH-LEVEL WASTE: GENERAL SUMMARY

High-level waste, in the form of either spent fuel rods or glass made from reprocessing waste, according to current plans will be disposed of in mined excavations in rock at depths of at least a few hundred meters. An excavation will remain open during the time required for emplacement of the waste (presumably a few decades) and for a short time thereafter, to permit retrieval of the waste if this should prove necessary, and then will be backfilled and sealed. Repository sites will be chosen at places where the rock at suitable depths is able to maintain openings for the necessary emplacement time, where ground water is slow-moving, where the path that ground water must follow to reach the surface is long, and where the rock is capable of precipitating or sorbing dissolved radionuclides that move through it. The geologic barriers to radionuclide movement will be supplemented by engineered barriers, to include at a minimum a highly insoluble waste form, a canister metal resistant to corrosion, and backfill material designed to slow ground-water movement, to sorb radionuclides, and to maintain a suitable chemical environment.

The behavior of waste in a repository sited and constructed according to these specifications can be predicted with reasonable assurance on the basis of known properties of the waste, the canister metal, the backfill, and the surrounding rock. The predictions are made from models based on scenarios, imagined sequences of events in which engineered barriers eventually fail and ground water comes into contact with the waste.

Different scenarios assume complete failure of the engineered barriers, complete failure of the geologic barriers, or partial failure of some components of either, at various times after repository closure. Some of the models are quantitative, giving numerical estimates, with calculated uncertainties for the amounts and concentrations of radionuclides that will reach the biosphere at different times and the resulting radiation doses to humans. Most of the models, except those for extreme scenarios, give assurance that the amounts escaping will be too small to pose a significant hazard to living creatures, and that the system has enough redundancy to compensate for possible failure of one or even several of its barriers. Comparison with the geologic record of the past long-term behavior of analogous nonradioactive elements provides a further indication that the models do not have gross errors.

This is the substance of the technical case that can be made in support of current plans for the mined geologic disposal of high-level nuclear waste. It is a strong enough case to convince most scientists and engineers that the technical problems of HLW disposal are well in hand.

Not everyone agrees that the situation is quite so rosy. Assumptions used in the models all involve uncertainties, the uncertainties are magnified as projections are extended farther and farther into the future, and focusing on the uncertainties can bring parts of the technical case into question. A skeptic can reasonably complain for example, that we don't know for sure whether the harmful effects of radiolysis of ground water would be always limited to the immediate vicinity of a repository, or how fast borosilicate glass will disintegrate under the combined influence of heat and radiation, or whether at a given site all possible paths of ground-water flow to the surface have been detected, or how safely retardation factors determined from short-term experiments in the laboratory can be extrapolated to the long-term behavior of fluids in rock. Some objections of this sort may be answerable by additional research, and practically all technical workers would agree that on at least a few points more research is indeed desirable.

But this leads to the difficult question of just how much research is essential. However long research may be continued, there will always be points that could be further studied. A skeptic can always find reasons for objection, reasons for demanding yet further research before repository construction can be approved. Where does one call a halt? How is the decision to be made that enough data are at hand, that further research is superfluous? This is partly a technical question, but in larger part a political one: a repository can be constructed only when society, through its elected representatives, has made a decision that the risks of proceeding without more research are acceptable.

Of the possible alternatives to mined geologic disposal, the most attractive is burial of HLW beneath the red clay that covers large areas of the deep-sea floor. Much can be said in favor of subsea-bed disposal, but the need for additional research is more apparent here than for disposal on land, and the political problems of decision making seem even more formidable.

9

Waste that is not high-level

9.1 LOW-LEVEL WASTE: PROBLEMS OF DEFINITION

If high-level waste by definition is restricted to spent fuel rods and the liquid and solid materials that can be directly derived from them, low-level waste (LLW) becomes a wastebasket term that includes all other varieties. The definition of LLW and its many kinds varies from country to country and also from time to time, as governmental bodies revise their regulations about managing different varieties of waste. Most LLW contains very little radioactivity. Some has substantial concentrations, and is often designated 'medium-level' (MLW) or 'intermediate-level' (ILW), but these terms have no generally accepted definition. In the United States two varieties of the non-high-level waste are singled out for special consideration: transuranic or TRU waste, which is material containing appreciable amounts of the heavy actinide elements (chiefly Np, Pu, Am), most of which is generated during the production and handling of plutonium for military purposes; and the waste or 'tailings' from the mining and milling of uranium ore, which has accumulated in large quantities at the many mill sites in the western part of the country. In Europe, medium-level waste is commonly distinguished from LLW on the basis of radioactive content, the definitions varying from country to country; and MLW is subdivided into varieties that do or do not contain transuranic elements.

Low-level waste proper (excluding mill tailings and TRU waste) comes from many sources and may contain a wide variety of radioactive elements. Besides that generated in reactor operations and the reprocessing of spent fuel, much of it comes from the use of radioactive materials in medical practice, in research laboratories, and in industrial processes. The radioisotopes in waste from these latter sources are commonly different from those prominent in reactor waste, and most have half-lives of no more than a few decades. In the United States about as much LLW comes from these sources as from the generation of nuclear power.

Material classified as LLW includes gases, liquids, and solids. Among the gases, ^{85}Kr and ^{133}Xe are fission products from ^{235}U; ^{14}C (as CO_2) and ^{3}H (tritium, in H_2O) come from secondary reactions in reactors. Large quantities of such radioactive gases have been allowed to escape into the atmosphere, where rapid dilution reduces their concentration to low values, but regulations in the United States now require that all except ^{133}Xe (whose half-life is only five days) be in large part trapped and disposed of with other waste. Liquid waste, most of it very dilute, includes huge volumes of cooling water and condensate from reactor and reprocessing operations, and smaller volumes of laboratory waste.

Examples of low-level solid waste are laboratory debris (towels, gloves, filter paper, glassware, broken and discarded apparatus), carcasses of animals used in biological experiments, the metal tubes of fuel rods after the spent fuel has been dissolved from them, and the discarded containers of radioisotopes used in medicine for treatment and diagnosis. Soil contaminated by disposal of liquids in the past, either accidentally or deliberately at a time when standards were less strict than they are today, is another possible kind of waste; but whether an attempt to exhume the contaminated material for disposal elsewhere entails more risk than leaving the soil in place remains an open question. Another troublesome kind of solid waste in the future will be the large chunks of structural debris resulting from the dismantling of reactors as they become obsolete and are replaced by new ones.

The radioactive material in most LLW consists largely of isotopes with short half-lives, and includes little or none of the heavy alpha-emitting isotopes with very long half-lives (atomic masses greater than that of lead). The rapid decay means that the isolation time required for such waste is a matter of a few centuries rather than the tens or hundreds of millennia needed for HLW. The concentration of radioactive material in much of the waste is so small (averaging less than 370 MBq/m^3) that it generates no appreciable heat and can be handled with little or no protective clothing. This does not mean that LLW is harmless: damage to tissues from long-term exposure is cumulative, so that isolation from surface environments is important for low-level waste just as it is for high-level varieties.

9.2 LOW-LEVEL WASTE: PROBLEMS OF DISPOSAL

The proper management of solid LLW is a subject of long-standing controversy. In the United States the common method of treatment has been shallow landfill, the placing of waste containers in shallow trenches and covering them with a meter or two of dirt (Figure 9.1). The obvious hazard in this procedure is the possible penetration of rainwater into the trenches and the movement of dissolved radionuclides into nearby streams or subsurface aquifers. Theoretically this can be largely prevented if landfill sites are located well above the permanent water table, in areas where surface drainage is good and underlying material has fairly low

Figure 9.1 A typical trench for shallow-landfill disposal of low-level nuclear waste. Metal boxes containing the waste are being stacked in the trench. When available space has been filled, earth and a protective covering for erosion control will be placed on top of the waste containers. (U.S. Department of Energy)

permeability. A site can often be improved by treating and shaping the cover to shed most of the water that falls on it. But the record of experience with landfill in the United States is not encouraging; some of the existing sites have functioned well, but enough have leaked small quantities of radioisotopes to raise questions about the advisability of continuing to use this method of waste management.

Shallow landfill is also an accepted disposal method for LLW in several European countries. MLW without transuranic elements is included with LLW at landfill sites in France, but not elsewhere; MLW with appreciable amounts of the long-lived elements, it is generally agreed, will need burial at deeper levels. Disposing of LLW by dumping it in the deep waters of the ocean was a common practice in Europe until 1982, and in the United States until 1970; many feel that this is a safe and efficient disposal method which should be resumed. Some European countries plan to dispose of LLW and MLW underground: in Sweden, in a repository excavated at shallow levels (50–150 m) under the floor of the Baltic Sea; in West Germany, in an abandoned iron mine; in Switzerland, in a tunnel driven horizontally into a mountainside. Such engineered alternatives of course, are considerably more expensive than landfill or ocean dumping.

A unique method of handling intermediate-level waste (up to 9250 MBq/m³) has been practiced for many years at the Oak Ridge National

Laboratory in Tennessee, and has been considered for use elsewhere. The waste, in liquid form, is mixed with cement and other solids to form a slurry that is injected at a depth of 200–300 m into a thick shale formation which has been previously cracked along bedding planes by water under high pressure ('hydrofracking', a technique widely used in oilfield practice to enhance oil recovery). The slurry sets to a solid after a few hours, and thereafter remains as an inert layer between shale strata. The technique requires care to ensure that the slurry has the right composition to set quickly; in 1985 one batch prepared in haste failed to solidify completely, and small amounts of radionuclides were detected in ground water near the site. Concentrations were far below regulatory standards, but state authorities used the leakage as an excuse to terminate the operation. Whether it will be resumed is uncertain. Proponents of the method feel that termination was not justified, and think that the procedure might be applicable at other sites with appropriate geology, possibly even for some varieties of high-level waste.

Because of the fuzzy definition of LLW, it is hard to give a firm figure for the magnitude of the disposal problem. In the United States the volume of waste is especially large because so much has been generated by the military program: in the disposal sites reserved for military LLW, a total volume of more than 2×10^6 m^3 had accumulated by the mid 1980s. The volume from commercial sources was less than half this figure, but if present rates of waste generation (estimated as some 100 000 m^3/yr) continue, the totals will soon be comparable. In Europe the amounts are somewhat smaller: estimated volumes of LLW to be disposed of by the early 1990s, including waste that is already buried, are between 400 000 and 700 000 m^3 for France, West Germany, and the United Kingdom.

9.3 TRU WASTE

Low-level waste containing appreciable amounts of the heavy actinide elements (called 'TRU waste' in the United States and 'MLW with alpha emitters' in Europe) needs special consideration. Waste with only traces of the heavy elements can be treated as ordinary LLW; in the United States the limit has been set at 3700 Bq/g. With concentrations beyond this limit, the waste requires isolation from the biosphere for the same long times as those needed for high-level material. One obvious way to dispose of such waste is simply to include it with the HLW that is destined for deep underground repositories, and this is the planned procedure in countries where the amounts are small. European countries with larger quantities are storing it for the present near the surface, with the intention of placing it ultimately either in an HLW repository or in a separate underground cavity at a comparable depth.

In the United States, where the amount of TRU waste is very great because of large-scale production and handling of plutonium, a special repository for this waste is under construction in the thick Permian salt beds that underlie much of south-eastern New Mexico (the Waste Isolation Pilot Plant, or WIPP). The plan is to bring TRU waste from the rest of the country to this one site. Although the waste requires isolation in a deep repository, the problems of managing it are simpler than for HLW because it generates little heat and because much of it has so little radioactive content that it can be handled without remote control. Concern about long-term isolation focuses on the possibility of ground-water intrusion, just as it does for HLW; but the concern is relatively minor, because here there will be no temperature rise to change the composition or structure of the surrounding salt and thus conceivably to provide access for water. At the WIPP site actual construction of a mined geologic repository is well advanced (Figure 9.2), and burial of waste is expected within a few years – in contrast to repositories for HLW, which are still only in the early

Figure 9.2 A continuous mining machine trimming the walls of a tunnel in rock salt at the WIPP site. Tunnels in a salt repository for high-level waste would be similar. (WIPP Operations Office, U.S. Department of Energy)

planning stages and will not be ready to receive waste for at least another two decades.

The WIPP facility will accommodate TRU waste that is currently stored in containers at the surface near the sites of production (totaling some 70 000 m³), plus the waste that will be produced for many years into the future. A more difficult question is raised by TRU waste that was generated in the early years of plutonium production. At that time it was thought that LLW containing plutonium could be adequately managed like other LLW, by shallow landfill – putting the waste, as either liquid or solid, into trenches and covering it with dirt. The plutonium, it was assumed, would be immobile even if water did percolate through the trenches, because of the insolubility of its compounds and its strong sorption on mineral surfaces in soil and rock. The assumption has proved to be generally correct, in that monitoring instruments at most disposal sites have shown only slight movement of plutonium in the three or four decades since it was buried. At a few sites however, the movement has been unexpectedly large, and in any event the existence of large volumes of soil containing low concentrations of plutonium (and other actinides) is by modern standards a cause for concern. Should the contaminated soil beneath the trenches now be exhumed and taken to WIPP or a similar facility for permanent disposal? The question is hard to answer. Certainly this material with its long-lived radioisotopes would be less of a hazard to future generations if it were placed deep underground; on the other hand, the isotopes show only a slight tendency to move in the present climate regime, and exhuming them would be both horrendously expensive and a serious hazard to operating personnel. According to present thinking, less harm will be done by leaving the soil in place than by trying to move it.

The WIPP repository will serve another function besides disposing of TRU waste: because it is similar to the HLW repositories that may be constructed later in salt beds, it provides a valuable opportunity for experiments, particularly for *in situ* testing of the response of a salt environment to the temporary presence of canisters containing simulated or actual high-level waste.

9.4 URANIUM MILL TAILINGS

Another kind of low-level waste that looms as a particularly serious problem in the United States is the debris from the mining and milling of uranium ore (Figure 9.3). The tailings are a major concern chiefly because of their enormous bulk, a total of some 140 million tonnes covering some 1500 hectares at 24 sites in the western states – largely a residue from the extensive mining activity of the 1950s and 1960s. The material is finely comminuted uranium ore, from which nearly all the uranium has been

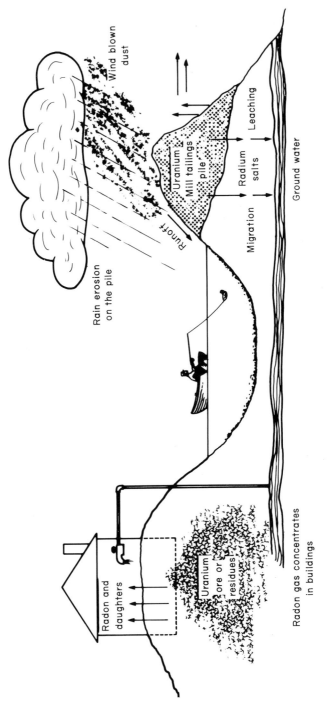

Figure 9.3 Releases of radioactive material from piles of debris produced by the mining and milling of uranium ore. (U.S. Department of Energy).

removed but which still contains the radioactive daughter elements that were produced by radioactive decay. Of these decay products the one posing the greatest hazard is an isotope of radium (^{226}Ra), which is dangerous for two reasons: it may itself cause damage if ingested, because of its intense alpha radiation; and its decay produces the radioactive gas radon (^{222}Rn), which can escape into the air and by further decay generate solid radioactive products (chiefly ^{210}Pb) as suspended particles, which on being inhaled can lodge in lung tissue and cause the sort of cellular damage that may lead to cancer. The decay products were of course present in the ore before it was mined and milled, but did no harm because they were covered by enough rock and soil to absorb the radiation and prevent appreciable movement. Now in the tailings piles they are at the surface and free to move, the radium by dissolving in water and the radon by diffusing into air.

In the early days of large-scale milling of uranium ore, the potential hazard from the accumulating piles of tailings was not realized. Material from some of the piles, in appearance a clean and easily accessible sand, was even used for construction of houses and school buildings. When the danger was recognized, replacement of parts or all of these structures was necessary, at enormous expense to the government. Now, of course, any usage of the tailings is prohibited and ample warnings are posted around the piles.

Control of radiation from the piles is possible by covering them with two or three meters of earth and landscaping to minimize infiltration of rain water. This is an expensive operation, but certainly warranted where the tailings are in or near inhabited areas. The even more expensive complete relocation of tailings piles has seemed necessary at some urban sites. Whether such drastic measures are justified for tailings remote from habitations remains a point of dispute. The amounts of radium and radon are not large, no more than a few hundred times the minute concentrations that are present in ordinary rock and soil; many think that the elaborate control measures are needlessly conservative, that some sort of restraint by fences and warning signs to keep the public away would be equally effective in minimizing exposure. The danger from the tailings is real enough, but whether it is sufficiently serious to warrant expenditure of several billions of dollars for complete elimination is one of the many questions about radioactive waste that can excite endless political debate.

Tailings are a problem wherever uranium has been mined and recovered from its ore in significant amounts, but in most countries the quantity is smaller than that resulting from the large-scale treatment of sedimentary ores in the western United States. The world-wide concern with this problem is indicated by a series of conferences held by the International

Atomic Energy Agency and by its publication of a brochure outlining safety standards and methods of tailings management.

9.5 LOW-LEVEL WASTE: SUMMARY

Low-level waste includes a variety of materials for which methods of isolation are far from uniform and often to some extent enmeshed in controversy. Uranium mine and mill tailings near inhabited areas can be rendered harmless by relocation or by using a sufficiently thick soil cover, but whether the expenditure needed for such operations is warranted for all tailings piles remains a matter of argument. TRU waste, it is generally agreed, requires isolation in deep mined repositories, either in those to be used for high-level waste or in repositories specially designed for this purpose. For other LLW, including some medium-level material, alternative methods of management include shallow landfill, ocean dumping, confinement in engineered structures at the surface, and underground repositories. Cheapest and easiest of these methods is dumping at sea, but to many people this is not a satisfactory alternative because of worry about contaminating the marine environment. How justified this concern may be is not certain; the amount of radioactive material is small compared with the huge volume of the ocean, and no demonstration of harm has resulted from the many years of ocean dumping before the practice was terminated. Of the other alternatives shallow landfill is certainly the least expensive, and experience at many sites has shown that with careful site selection and good engineering oversight effective isolation of radionuclides can be achieved. A few conspicuous failures and increasing doubts about the long-term efficacy of this method have led some to question the advisability of its continued widespread use. The balancing of cost against risk and benefits seems especially difficult for low-level waste.

Institutional aspects of waste disposal

10.1 THE POLITICS OF REPOSITORY SITING

From a technological viewpoint the disposal of radioactive waste is a difficult but manageable problem. Even for the worst part of the problem, the disposal of high-level waste, the steps leading to safe isolation seem straightforward. The purely mechanical operations required – the sinking of a shaft and excavation of a repository area, the enclosing of waste in canisters and movement of canisters into the repository, backfilling the repository and sealing the openings – can all be accomplished with known engineering techniques. Geological criteria for choosing repository sites are well established, and a number of places that satisfy these criteria have been found. Some questions remain about details, particularly with regard to predictions of repository performance over the long-term future, and some of the details require further study. Among scientists and engineers who are familiar with the problem there is fairly substantial agreement that the needed research is minor and can be accomplished as the construction of a repository proceeds. Why then, do we still delay? Why isn't a repository for HLW already under construction?

There are many answers, partly technical and partly political. On the technical side, some knowledgeable and vocal scientists still have misgivings. They feel the need of more assurance that we have found a satisfactory form for the waste, the best metal for the canisters, and an appropriate material for backfill, plus more in-depth study of the favored repository sites, before a decision to go ahead should be made. The consequences of an error, they point out, could be catastrophic, and rectifying an error would be prodigiously expensive. For some of the objectors a major consideration is doubt that spent fuel should ever be put underground: building a repository should be deferred, they say, until bureaucrats come to their senses and encourage the reprocessing of fuel to recover its potentially energy-producing isotopes. A few would go further, as we have noted before, maintaining that to put any waste underground is a mistake, because it contains a variety of unusual metals which may become important for future technological processes whose nature we cannot predict. A wiser course, in their view, would be to store all radioactive waste in an accessible form at or near the surface, surrounded by enough natural or artificial bulwarks to keep radiation adequately controlled. Still others emphasize the advantages of delay, on the grounds that the heat and radiation emitted by HLW decrease with time, and hence that disposal will be easier to accomplish several decades from now than it is today. Faced with these conflicting opinions among technical experts, it is small wonder that the non-experts who must participate in the ultimate decision to start repository construction are bewildered.

But the main reason for delay comes from another source, the unwillingness of most people to have a disposal site located in their vicinity. Despite all the assurances that government agencies can give about the lack of danger from a repository, the fear of radioactivity is so strong and so widespread that a negative response among the local population is immediate whenever a particular area is suggested as a possible repository site. Even attempts at preliminary geologic study of an area are sometimes frustrated by local opposition. Almost everyone agrees that waste disposal is necessary, but almost no one wants the disposal operation to take place anywhere near his or her domicile.

The objections provide good political capital for local office-holders, who can pose as defenders of their constituents' rights against the designs of a distant bureaucracy. The strength of popular objection varies greatly of course, from one country to another and among population groups within each country. A few communities even welcome the prospect of a disposal operation nearby, because of the boost to the local economy from jobs that the project would create. But in general this conflict between an agency of big government that is trying to find a repository site and citizen groups who feel imposed upon looms as a major hurdle and a major cause of delay in getting repository construction actually started.

Not only politicians but environmental organizations take advantage of local anti-repository sentiment. Many of these groups are sincere in their efforts to protect environmental quality, but some have ulterior motives – for example, a hope that by delaying repository construction indefinitely they can show that safe disposal of radioactive waste is impossible, hence that generation of more waste should be prohibited and the nuclear-power industry should be shut down. Whatever their motive, some environmental groups are adept at maintaining prolonged litigation that can complicate or frustrate the choosing and developing of repository sites.

When the time comes, a few years or a few decades hence, for actually settling on a specific site or sites and getting construction under way, how can such local opposition be appeased? The obvious answer is in part education and in part a determined effort by government agencies to work with local groups. Education of the public on a subject as complicated as nuclear-waste disposal is necessarily slow, but a sure sign of progress is the fact that objections to the search for repository sites are less strident than they were a few years ago. Also in recent years government representatives have shown increased willingness to consult concerned citizen groups, at times to even subsidize their activities. The problems of communication are by no means entirely solved, but much progress has been made. A further measure often suggested for softening local objections is the offering of some kind of quid pro quo, say a tax reduction or improvement of public works, as compensation to a community for willingness to accept the onerous burden of a nearby repository.

10.2 NATIONAL INSTITUTIONAL ARRANGEMENTS

Responsibility for the safe disposal of nuclear waste, it is generally agreed, rests with national governments. Different countries have widely different organizational patterns for accomplishing the varied tasks necessary for waste disposal – setting standards, finding sites, conducting research, preparing the waste, excavating repositories, emplacing the waste, sealing the repositories. In some countries all such tasks are performed, or will be performed, by government agencies. More commonly the agencies have a supervisory and regulatory role, and the actual work of planning, constructing, and filling repositories is to be accomplished by private concerns under contract. The private concerns are often the major utilities that generate nuclear power and produce waste as a by-product, or combinations of such utilities; a common pattern is quasi-public corporations that include representatives from both government and the nuclear-power industry. The necessary research may be done by government laboratories, universities, utilities, or consulting firms, or by all of these. Because the United States, among western countries, has the largest quantity of waste requiring disposal and the most complex administrative set-up, a brief glance at its institutional arrangements will serve as an example.

Three agencies play major roles in the United States waste-disposal enterprise: the Environmental Protection Agency (EPA), which sets general standards for the permissible release of radionuclides from a repository; the Nuclear Regulatory Commission (NRC), which establishes rules for the siting and construction of a repository so as to meet the EPA standards; and the Department of Energy (DOE), which has the responsibility for finding sites and seeing that repositories are built. Actual construction and repository operations are delegated to private contractors. Other federal agencies have peripheral functions; for example, the

Department of Defense supervises management of wastes from plutonium production until they are turned over to the DOE for disposal, and the Department of Transportation is concerned with safety in the movement of waste to repository sites.

All the agencies work in response to legislation enacted by the U.S. Congress. The most recent law regarding disposal of high-level waste is the Nuclear Waste Policy Act, signed by the president in 1983 and amended in 1987, which spells out in detail the steps that must be followed by the DOE in locating sites, studying them thoroughly, choosing a small number from which the president is to select one for the first repository, and seeking a construction license from the NRC – all with the mandated goal of having the first repository ready to receive waste in 1998. The act specifies also that the large sums needed to carry out the project will be supplied by a small levy on each kilowatt-hour of electrical energy produced by the utilities that generate the commercial waste, which of course will be passed on to consumers of the energy, the intention being that those who benefit from the production of nuclear power should pay for disposal of its waste. The rigid schedule laid out in the Act is proving hard for the DOE to meet, and a recent request for modification of the Act would defer opening of the first repository until 2003. While the DOE struggles to maintain its assigned schedule, it is also supporting a myriad of research projects on various aspects of waste disposal, in national laboratories and universities throughout the country. The Act outlines further the procedures that the DOE must follow in cooperating with states, Indian tribes, and local communities so as to avoid or mitigate the conflicts between federal and local interests mentioned above.

Questions are often asked as to whether this ponderous procedure, with three large agencies involved and Congress continually looking over their shoulders, is the most efficient way to solve the disposal problem. Inevitably differences of opinion crop up among the agencies, and smooth operation has often been hampered by overly rapid changes in personnel. A quasi-public corporation, with broad powers to get on with the job quickly, has been suggested as a more efficient alternative. More efficient it might well be, but whether it would be sufficiently sensitive to the feelings and wishes of the many individuals and groups who are in one way or another affected by decisions about waste management is not certain. Progress with the EPA–NRC–DOE combination at times seems discouragingly slow, but over the years progress is indeed perceptible and the interests of many diverse parties have been well protected.

In contrast to the management of HLW, most low-level material is left in the hands of the individual states or regional groupings of states. This is not true of transuranic waste, which is a DOE responsibility, and the DOE is also charged with stabilization of the tailings piles at abandoned uranium mining and milling sites. State responsibility for LLW has only recently been spelled out by Congress, and how successful this management plan will be is not yet clear.

In France, West Germany, and the United Kingdom institutional mechanisms for the management of nuclear waste are similar to the United States pattern in that government entities have been established to supervise disposal operations and to license disposal facilities, while standards for limiting radioactive release are set by other agencies whose primary concern is public health or the environment. Similar also is the financing of disposal activities by fees paid by the energy-producing utilities, hence ultimately by the users of the energy of which the waste is a by-product. Organizational patterns differ in the amount of direct involvement of utilities in the waste-management agencies, in the contracting of disposal operations and research to private concerns, and in the assignment of responsibility for managing LLW and MLW as well as HLW to a single agency or more than one. In countries with a small nuclear energy program the institutional arrangements can be simpler, but all include in one way or another the three functions of waste management, licensing of facilities, and setting standards for permissible release.

10.3 INTERNATIONAL ORGANIZATIONS

The safe disposal of nuclear waste is a problem faced by all nations in which radioactive materials are produced or used, and international cooperation in solving the problem has long seemed imperative. Radioactive wastes have no respect for political boundaries: ground water containing dissolved radionuclides and windblown dust contaminated with plutonium can be hazardous far beyond their country of origin. To ensure that all nations with waste-disposal programs are kept aware of current thinking about the management and safe isolation of different kinds of waste requires the frequent holding of multi-national conferences and the wide dissemination of research results. For such purposes several international organizations have been set up.

Most inclusive is the International Atomic Energy Agency (IAEA), which numbers among its members more than 80 countries in all parts of the world (China and Taiwan are notable exceptions). Waste disposal is only one of the concerns of the IAEA, but in this field it has been active in organizing international meetings and symposia, convening technical committees and advisory groups for particular topics, holding training courses, and coordinating research programs. A smaller group of industrial nations makes up the Nuclear Energy Agency (NEA) of the Organization of Economic Cooperation and Development (OECD); originally restricted to western Europe, the NEA now includes Australia, Canada, Japan, and the United States. Like the IAEA, and often in cooperation with it, the NEA has held conferences and workshops on waste management and has sought to initiate and coordinate multi-national research projects. An organization similar to OECD among eastern-bloc nations is the Council for Mutual Economic Assistance, which has likewise set up an agency primarily concerned with problems of nuclear waste management. Other smaller groups of European nations have established agencies that deal with specific questions of waste management and promote joint research on problems of mutual interest.

To consider the special problems of the subsea-bed disposal alternative, an international organization is particularly important. Such an organization was established in 1974, as a subgroup of the NEA; its membership includes most of the nations with a major interest in this possible disposal method. The selection of large tracts of the ocean floor that would be particularly suitable for further exploration is one accomplishment of this group.

Among the multi-national research projects that have been active during the past decade, those carried out in the underground laboratory of Stripa in Sweden are outstanding examples. As mentioned in Chapter 6, the Stripa facility consists of several tunnels that have been opened in a granite body adjacent to a worked-out iron mine, providing an opportunity for extensive *in situ* experiments on the behavior of ground water and of waste-package materials in a crystalline rock environment. Some of the experiments are performed by Swedish scientists, but many are proposed, carried out, and partly financed by workers from other countries (West Germany, United Kingdom, Switzerland, United States). Much of our present knowledge about the effect of heat on fracturing and water movement in granite and on the properties of bentonite as possible backfill material has come from this one laboratory. Similar underground facilities in crystalline rock have been set up recently at Grimsel in Switzerland and at Pinawa in Manitoba, with the intention of inviting research projects from other countries. Multi-national research on waste disposal in salt is under way in a disused salt mine at Asse in West Germany, and on disposal in clay at Mol in Belgium.

The multiplicity of international agencies would seemingly lead to redundant efforts and possible friction. In general such troubles have not materialized, the agencies have worked well together, and their record of accomplishment is excellent. The accomplishments, it should be noted, are entirely in facilitating joint discussion and research, and not in the actual handling of waste; earlier proposals for the development of common disposal sites to be used by several countries did not bear fruit, and at present each nation is assumed to be responsible for disposal of its own waste. But for tackling the general background problems of waste management, the need for international cooperation is so manifest that national rivalries have played only a minor role.

Some questions of policy

11.1 INTRODUCTION

Despite widespread agreement that mined geologic repositories are the best solution to the problem of waste disposal, some in both the technical community and among the general public remain skeptical. We noted some of the grounds for skepticism in earlier discussions. Because the skepticism raises important questions of general policy, it deserves fuller examination in this final chapter. Most of the questions straddle the border between technology and socio-political considerations; as with many other problems in the modern world, the public and its elected representatives are asked to make decisions about matters of emotional import where there is some uncertainty in technical knowledge and some disagreement among the technical experts themselves. The problem of nuclear waste disposal has the added difficulty that it involves predictions about times in the far distant future.

We pick five questions of policy for discussion: (1) Should HLW be maintained in near-surface storage, rather than buried underground? (2) Should spent fuel rods be reprocessed to recover their contained fissile elements? (3) How effectively must HLW be isolated from the human environment, and for how long is isolation needed? (4) Should every effort be made to dispose of HLW as soon as possible, or should disposal be deferred? (5) How much additional research is required before actual construction of a repository can begin? All of these topics have been touched on briefly in earlier chapters.

11.2 SURFACE STORAGE VERSUS UNDERGROUND DISPOSAL

The feasibility of storing HLW safely at the earth's surface for considerable periods is amply demonstrated by the satisfactory performance of the water basins containing spent fuel rods at reactor sites and the steel tanks filled with waste from reprocessing at installations for plutonium production. Such storage devices require constant surveillance; the need for long-term continuation of this surveillance is the chief argument for resorting to a more elaborate and more permanent method of disposal. The water basins and steel tanks were designed only as stopgap measures, temporary structures to hold the waste until permanent repositories could be made ready to receive it. But certainly it should be possible to build more substantial protective structures for the waste at the surface, which would adequately shield the surroundings from radiation and which would require little or no human attention for a very long time. The pyramids of Egypt come to mind as possible examples, although surely structures of smaller size and more efficient shape could be devised. Alternatively tunnels could be driven into the impermeable rock of a mountainside, with their entrances blocked after the waste was emplaced. Either a massive edifice at the surface or a mountain tunnel would need no surveillance beyond a possible occasional check of monitoring devices, and either would permit fairly easy access to the waste if this should ever seem desirable in the future.

Only a minimum of geologic precautions would be needed for siting such surface repositories. Regions of frequent earthquakes should be avoided, as well as regions where rapid erosion by water or ice can be expected. For the tunnel option, ground water would be a possible concern; but locating the tunnels in a region of arid climate and constructing channels to divert occasional ground-water flow away from the waste should take care of this worry. Siting problems would be minor compared with those for a deep mined repository.

In favor of surface storage are two strong arguments. First is the ease of retrievability: if something goes wrong and monitoring devices indicate hazardous escape of radionuclides, the waste could be readily recovered and moved to a safer place – much more easily than from a sealed repository several hundred meters underground. The second argument is similar but rests on the assumption that some use will be found for the waste in the future: if a way is found to utilize the energy or the radiation generated by HLW, or to apply some of its material to metallurgical purposes, the waste would again be readily accessible. This possibility could be very real of course, if the waste consists of spent fuel, because its contained ^{235}U and ^{239}Pu might well be attractive as a future energy source.

The opposing arguments are pretty obvious. Waste at or near the surface, no matter how well protected, is always more vulnerable to exposure than waste underground, either from natural causes or from deliberate or inadvertent human intrusion. A structure on the surface, even one as massive as the pyramid of Cheops, is not immune to weathering and erosion over tens of thousands of years; a cavern in a mountainside would be prey to landslides as well as rapid stream erosion. Intrusion by humans is especially likely if the waste includes spent fuel, since the plutonium it contains would be a target not only for those seeking an energy source but also for would-be trouble-makers hoping to fabricate a bomb.

The opponents of permanent surface storage at present have the better of the argument, but the advantages of keeping the waste where it can be watched and where it is readily accessible if needed have strong appeal.

11.3 TO REPROCESS OR NOT TO REPROCESS?

From the standpoint of energy conservation, there is no question that spent fuel should be reprocessed to recover its fissile material. If fuel rods are disposed of as waste, about a third of the usable energy content of the original uranium is thrown away. The reprocessing operation – dissolving the contents of the fuel rods, adding organic reagents to the nitric acid solution to collect the uranium and plutonium, fabricating new fuel rods from these elements, and converting the radioisotopes remaining in the acid solution to borosilicate glass – is a complicated procedure, not economically justified when the price of uranium is low. But it seems certain that eventually the world will need this supplemental source of energy. Reprocessing has the additional advantage that all HLW would then have a single form, borosilicate glass, and operation of a repository could be simplified by adaptation to this one kind of waste material.

The opposing argument rests on the simple fact that the fissile materials remaining in spent fuel can be used for making nuclear bombs. Making bombs would not be easy, because the combination of isotopes produced by reprocessing of spent fuel is not appropriate for an efficient bomb without additional isotope separation, but a determined group of terrorists could doubtless find a way to make a usable destructive device. If reprocessing were to become a common industrial operation, so runs the argument, plutonium would be produced in considerable quantity at many places and would be difficult to protect from pilferage.

The validity of this argument can be endlessly debated. As long as reprocessing is limited to a very few places under the close supervision of responsible governments, as it is today in France and the United Kingdom, adequate protection can be assured. If the price of uranium increases, as it very likely will in the near future, reprocessing will be more economically attractive and may become more widespread. Whether sufficient protection from would-be bomb makers can be provided if plutonium recovery proliferates is a major problem for the world's political leaders.

Reprocessing of commercial nuclear fuel is currently practiced on a modest scale in a few countries, but the fear of terrorism, coupled with lack of economic incentive, has so far prevented its general adoption as a routine procedure. Thus the disposal of unreprocessed spent fuel remains a major part of present plans for management of HLW in most countries.

11.4 WASTE ISOLATION: HOW EFFECTIVE, AND FOR HOW LONG?

No geologic repository, it is generally agreed, can keep radioactive waste completely out of surface environments for all future time. Constructing a repository, after all, is a human enterprise, an enterprise that will change a portion of the natural environment profoundly. Any such enterprise can be expected to fail eventually, either from natural causes or from defects in human workmanship. Normally we count on engineered structures like dams or bridges or large buildings to serve their purpose for decades or perhaps a few centuries, but eventually they fall into disrepair and must be modified or replaced. As an extreme, examples can be cited of structures from the ancient world that by accident of benign climate and good engineering have survived reasonably intact for a few millennia. Now for waste disposal we have the effrontery to plan underground structures that are expected to endure not for thousands of years but for tens or hundreds of thousands. Is this reasonable? How long can we expect these structures to serve their purpose of isolating waste? What is the minimum time that complete containment is needed? Thereafter, what degree of failure can be tolerated? How can this minimum failure be achieved, and the achievement guaranteed? Questions like these are among the most difficult that disposers of waste must face, and it is not surprising that complete agreement about answers has not been reached.

The difficulty goes back to the impossibility of specifying the physiological damage that results from very low levels of radiation. If we knew that radiation doses above a certain level could be harmful to some persons and radiation below this level was harmless to all, we could build repositories to this standard. Whether such a threshold exists we do not know, and will probably never know, because the statistical evidence on which it would be based has too much 'noise'. For the sake of caution it is commonly assumed that the threshold does not exist, that any radiation dose, no matter how small, is potentially harmful. This leads to a seeming contradiction, because we are continually exposed to natural ionizing radiation which causes no perceptible damage. One can argue of course, that natural radiation is actually harmful, that we would live longer and more happily in a radiation-free world – an impossible fantasy, because our bodies could not function without elements that have radioactive isotopes. Faced with this curious uncertainty, we have no choice but to adopt the pragmatic viewpoint that natural radiation, whatever harm it may do, is something that we (and all other living creatures) have learned to live with and can disregard. If doses of radiation from nuclear waste can be kept within the range of variation in the natural background, we can reasonably consider it harmless. It is this unsatisfactory standard that is commonly used as a measure of the effectiveness of waste isolation, but acceptance is far from universal. Many consider the standard needlessly low, while a few insist that any addition to the natural background is unacceptable.

So the answer to our question about the length of time needed for waste isolation must be an elastic one. Total containment is desirable for as long as is reasonable economically, and certainly techniques and materials are available to extending the time to at least a few centuries. After the containment period, repository design commonly calls for keeping the escape of radionuclides below some minimum value. The value should be related to the normal range of doses from natural radiation, but the best way to express it is a point of debate: should it be the minimum total escape of radionuclides into the environment, the minimum concentration in water moving out of a repository, or the minimum radiation dose to the maximally exposed individual? The rate of escape, however it may be expressed, can be predicted from plausible models; the models show that for a well sited and well constructed repository the safe minimum will not be exceeded at any time in the future, but predictions for times after 10 000 years become increasingly undependable because of uncertainty about geologic processes. HLW contains enough long-lived radionuclides that on complete exposure the radiation would be well above natural levels for many tens of millions of years, but complete exposure is unlikely. Thus the length of time for waste isolation cannot be specified as a number of years, but must be described in roundabout language: ideally the time should be of the order of a hundred million years, but this is absurd from both an engineering and a geologic viewpoint; complete containment should be required until the short-lived isotopes (especially ^{90}Sr and ^{137}Cs) have decayed to innocuous levels, say at least a few hundred years; and release of radionuclides thereafter should be kept to a specified low level for at least 10 000 years, with a strong probability that the time will be much longer. Such language is not the kind with which policy makers normally deal, and their discomfort is heightened by disagreement among experts as to both the times and the precise meaning of 'low level'.

11.5 DISPOSAL IN THE NEAR FUTURE VERSUS DELAY

Because disposal of HLW is widely perceived as a difficult technical problem – in some circles, as a well-nigh insoluble problem – there is much public demand for a convincing demonstration that adequate isolation can indeed be achieved. There is also a commonly expressed sentiment that this generation should demonstrate its ability to take care of its own nuclear waste, rather than leave the problem to our children and grandchildren. These are powerful arguments for getting on with waste disposal as quickly as possible. They underlie for example, the urgency implicit in the rapid schedule for repository construction mandated by the Nuclear Waste Policy Act passed by the U.S. Congress in 1982, and the earlier legislation in Sweden requiring demonstration of disposal feasibility before fuel could be loaded into just-completed nuclear reactors.

On the other hand, as noted earlier, a convincing technological case can be made for postponing disposal for several decades. All HLW at present is effectively isolated from the environment, and will remain so as long as the water pools containing spent fuel roads and the tanks of waste from reprocessing are maintained. The maintenance is not burdensome and can be continued indefinitely, so there is no technical urgency for doing anything at all with the waste. As the waste stands it becomes progressively easier to manage, because radioactive decay leads to decrease in both its temperature and the amount of radiation emitted. Furthermore, the advance of technical knowledge as time goes on will almost certainly lead to improvement in waste disposal techniques. Relaxation of the disposal schedule would provide more time for laboratory and *in situ* experiments, giving a firmer basis for geologic predictions about long-term changes in the repository environment and the long-term behavior of any radio-nuclides that escape. One can speculate also that progress in medicine may make the effects of radiation less fearsome – ways may be found to counteract cellular damage, or to reverse the growth of cancers – so that sometime in the future the stringent limits on allowable radionuclide release may be eased. For all these reasons many in the technical community think it would make sense to slow the haste with which disposal problems are at present being attacked.

Such reasoning underlies the explicit intention in Sweden to keep waste in surface storage for at least 40 years, and in England and France for 50 years, before underground disposal is attempted. In other countries it seems likely that a similar delay will be achieved quite unintentionally, because the decision to actually start repository construction will almost certainly be long put off by the ponderous working of government bureaucracies and the delaying tactics of local groups at sites selected for disposal operations.

Postponement of repository construction, whether intentional or not, can lead to another political complication. The water basins for spent fuel rods at reactor sites, although not difficult to maintain, are still an expensive nuisance for the utilities that must manage them, especially if they must be continually enlarged and watched over for many decades. The manage-ment of spent fuel could be more efficient, and the utilities could be relieved of an unwelcome burden, if the fuel rods were removed to a central storage facility where they could be kept under surveillance ('monitored retrievable storage') until a repository is at last ready to receive them.

Where should such a large central facility be located? This raises at once the same sort of problems that plague the siting of a permanent underground repository. Can a place be found where the geology will ensure containment of the waste in case of accident? And how can the fears of nearby residents be assuaged when they learn that large quantities of highly radioactive material will be kept for a long period in their vicinity? Such fears will be aggravated of course, by the thought that the facility for temporary storage might well become a de facto permanent repository if the siting and construction of the intended underground facilities elsewhere are indefinitely delayed.

One can have great sympathy for a politician faced with immediate decisions about deferring the disposal of HLW and allocating funds for construction of a temporary storage facility, truly caught in the horns of a complex dilemma – pulled one way by technical experts counseling delay, another way by members of his or her constituency eager for action in getting rid of an environmental menace, and still a third way by the unfortunates who feel themselves threatened by the prospect of having either a storage facility or a deep repository constructed near their homes.

11.6 IS MORE RESEARCH NEEDED?

Research over the past few decades in universities and government laboratories, plus field work at possible repository sites, has given us the fund of information on which plans for disposal rest. A truly impressive mass of data on all aspects of waste handling and repository construction has been assembled. Nevertheless a few details remain about which additional data would be useful. It would be nice to know for example, just how oxidizing we can expect ground water to become under the influence of radiolysis with different kinds of waste and in different kinds of rock environments; or how much the concentration of radionuclides in moving ground water is affected by diffusion and dispersion; or how fast intercrystalline water moves in salt under a thermal gradient; or how much the pH of ground water is affected by radiation damage to salt crystals; or how fast ground water is moving in the different aquifers at specific disposal sites. Research on such details is active at present, and certainly some of the research projects when completed will give us pointers on how disposal techniques can be improved. On the other hand, research is expensive, often prodigiously so. It seems fair to ask the question: is all this research really necessary? Are the added details important enough to be worth the expense?

Obviously research can go on forever – there will always be little points that need clearing up or numbers that could be made more precise. Conceivably the added information will make disposal safer. But a time must come eventually when the additional cost is perceived as not worth the modicum of safety gained. Many would say that the time has already come, that the research enterprise is developing a life of its own which will be hard to stop because research contracts are lucrative for many individuals and organizations. Other experts disagree, averring that precipitate repository construction would be foolhardy until many ongoing research projects are completed. Such dissension among authorities means a quandary for the general public and its elected representatives who must make the decision as to when repository construction should have a green light.

Basically the question is a matter of willingness to accept risk. It is easy enough to show, as has been emphasized repeatedly in other chapters, that the chance of unacceptable escape of radionuclides from an underground repository well located geologically and constructed with state-of-the-art engineering know-how is very small. Nevertheless the chance exists. There is always a finite risk that something will go wrong, no matter how carefully the geology of a site has been examined and how well waste packages have been prepared and emplaced. The risk can be lowered by further research, but it can never be reduced to zero. The risk from a repository built with present knowledge is almost certainly less than other risks in ordinary life that we accept without question – the risk of lighting a cigarette for example, or driving a car, or climbing a mountain. Are we, as a society, ready to accept this risk? How much money are we willing to spend on research to make the risk a little smaller? Are the benefits gained from the production of waste – the cheap and abundant electric power, the more dubious benefit of a huge arsenal of nuclear weapons – sufficient compensation for the risk involved in its disposal? If so, how much risk are the benefits worth? These of course, are political questions, and presumably they will be answered by politicians or by those whom politicians appoint to administer the responsible agencies. But surely the answers must rest ultimately on some sort of consensus among the general public as to the balance between risk and benefits and as to the magnitude of this perceived risk relative to others.

Beyond these matters of policy is a troublesome philosophical question relating to intergenerational equity. The question can be asked in many ways, all referring back to the simple fact that we of this generation are enjoying the benefits of cheap electricity and are thereby creating a risk and a hazard for generations to come. If a decision is made to defer waste disposal for a few decades, we will be presenting our decendants not only with a serious hazard but with a tough technical problem. Is it right for one generation to impose on another in this fashion? Do we have a moral obligation to dispose of our own waste as quickly as possible, despite technical reasons for delay, so that the burden will not be passed on to others? How does one balance costs against benefits, when the benefits are to us and the costs to our descendants?

One can rationalize our seeming selfishness in many ways. It is easy to point out for example, that preceding generations have shown little concern for the sort of world they were leaving to us, so perhaps our worry about the welfare of future generations is unwarranted. In the same vein, one can reflect that technological progress is rapid and unforeseeable, that we can hardly surmise what problems will seem important in the future world, that disposal of waste and existence of buried waste may seem trivial matters to our descendants, hence that we had best be concerned about our own welfare and leave our progeny to take care of themselves. Or we can note that production of waste is only one of many ways in which we are altering our planetary environment, and that future generations may regard this as a minor sin of our generation in comparison with our profligate wasting of the world's petroleum resources and our destruction of tropical rain forests. In a more egotistical mood, we can reflect on all the ways in which our generation has improved conditions of life – the greater ease of travel and communication, the advances in medicine, the new crop varieties, the conquest of space – and muse that people of the future should regard the waste disposal problem as a small price to pay for all the benefits we are bequeathing them. By such reasoning we can justify or trivialize our seeming unconcern with the welfare of those who will follow us, but the fact remains that we have indeed placed a burden on future generations. By producing intensely radioactive material from weakly radioactive ore we have (temporarily) increased the amount of radionuclides in the outer part of our planet; the nuclides are a problem to dispose of, and no matter how the disposal is accomplished they will be a potential menace to living things for many generations to come. It is important that we plan well, so that the menace will be as small as we can make it.

Further reading

For additional reading on various aspects of nuclear waste disposal, a vast literature is available. The following is a sampling of books and papers that give more detailed discussion, at roughly the same technical level, of some of the subjects touched on in this book.

General

Chapman, N.A. and McKinley, I.G. (1987) *The Geological Disposal of Nuclear Waste,* Wiley, Chichester pp. 280.

Milnes, A.G. (1985) *Geology and Radwaste,* Academic Press, London pp. 328.

Office of Technology Assessment, US Congress (1985) *Managing the Nation's Commercial High-Level Radioactive Waste,* US Government Printing Office, Washington DC, pp. 348.

US Geological Survey Circular 779 (1978) Geologic Disposal of High-Level Radioactive Wastes – Earth-Science Perspectives, *US Geol. Surv.,* Washington DC, pp. 15.

Standards for low-level radiation

Eisenbud, M. (1982) Radioactive waste management in *Outlook for Science and Technology: the Next Five Years,* W.H. Freeman, San Francisco, 287–320.

Geological criteria

National Research Council (1978) *Geological Criteria for Repositories for High-Level Radioactive Wastes: Technical Criteria,* National Academy Press, Washington DC, pp. 19.

Models

Bengtsson, A., Magnuson, M., Neretnieks, I. and Rasmussen, A. (1983) *Model Calculations of the Migration of Radionuclides from Spent Nuclear Fuel,* (KBS TR 83–48) Svensk Kärnbränslehantering AB/KBS, Stockholm, pp. 104.

Burkholder, H.C. (1984) Engineered components for radioactive waste disposal systems in *Radioactive Waste Management, vol. 4: Proceedings of an International Conference of IAEA, Seattle, 16–20 May 1983,* International Atomic Energy Agency, Vienna, 161–80.

National Research Council (1983) *A Study of the Isolation System for Geologic Disposal of Radioactive Wastes,* National Academy Press, Washington DC, pp. 345.

Disposal in salt, crystalline rock, and tuff

National Research Council (1970) *Disposal of Solid Radioactive Wastes in Bedded Salt Deposits,* National Academy Press, Washington DC, pp. 28.

Ubbes, W.F. and Duguid, J.O. (1985) *Geotechnical Assessment and Instrumentation Needs for Isolation of Nuclear Waste in Crystalline Rocks* (BMI/OCRD-24), Battelle Memorial Institute, Columbus, Ohio, pp. 185.

Svensk Kärnbränslehantering AB (1983). *Final Storage of Spent Nuclear Fuel, KBS-3, Summary,* SKBF/KBS, Stockholm, pp. 58.

Bedinger, M.S., Sargent, K.A., Brady, B.T. (1985) *Geologic and Hydrologic Characteristics and Evaluation of the Basin and Range Province Relative to the Disposal of High-Level Radioactive Waste* (US Geological Survey Circular 904-C), US Government Printing Office, Washington DC, pp. 27.

Natural analogs

Chapman, N.A., McKinley, I.G. and Smellie, J. (1984) *The Potential of Natural Analogues in Assessing Systems for Deep Disposal of High-Level Radioactive Waste* (NAGRA 84-41), Nationale Genossenschaft für die Lagerung Radioaktiver Abfälle, Baden, pp. 103.

Cowan, G.A. (1976) A natural fission reactor *Sci. American* **235,** 36–47.

Jakubick, A.T. and Church, W. (1986) *Oklo Natural Reactors: Geological and Geochemical Conditions – a Review* (Research Rpt. INFO-0179), Atomic Energy Control Board, Ottawa, pp. 53.

Low-level Waste

Atomic Industrial Forum (1986) *Low-Level Radioactive Waste: Building a Perspective,* Atomic Industrial Forum, Bethesda, Maryland, pp. 36.

National Research Council (1976) *The Shallow Land Burial of Low-Level Radioactively Contaminated Solid Waste,* National Academy Press, Washington DC, pp. 150.

Uranium mill tailings

National Research Council (1986) *Scientific Basis for Risk Assessment and Management of Uranium Mill Tailings,* National Academy Press, Washington DC, pp. 246.

Subsea-bed disposal

Hollister, O.D., Anderson, D.R. and Heath, G.R. (1981) Subsea-bed disposal of nuclear wastes, *Science* **213,** 1321–6.

Politics

Carter, L.J. (1987) Nuclear imperatives and public trust: dealing with radioactive waste. Issues in Science and Technology, **III,** no. 2, 46–61.

Further data about nuclear waste management may be sought in technical books, standard scientific journals, and symposium volumes, but by far the largest amount is in the so-called 'gray literature' – publications of national laboratories and consulting groups, published quickly for the benefit of other laboratories and government agencies, generally subject to less thorough review than articles in refereed journals. Convenient references to the more important items, both in standard journals and in the 'gray literature', are the extensive bibliographies in the books by Milnes and by Chapman and McKinley at the head of the above list. A more complete listing is contained in the abstract volumes ('Waste Management Research Abstracts') published annually by the International Atomic Energy Agency (see address below). Much of the 'gray literature' is not available in libraries, but indexes and individual items can be obtained from document repositories in the following list:

International
International Atomic Energy Agency (IAEA)
Waste Management Section
Wagramerstrasse 5
PO Box 100
A-1400 Vienna, Austria

OECD Publications Office
2 rue André-Pascal
75775 Paris CEDEX 16
France

Canada
Atomic Energy of Canada Ltd
Whiteshell Nuclear Research Establishment
Pinawa, Manitoba ROE 1LO
Canada

France
Commissariat à l'Energie Atomique DPS/SEAPS
BP No. 6
92265 Fonteney-aux-Roses
France

Germany, Federal Republic of
Kernforschungszentrum Karlsruhe GmbH
Institut für Nuklear Entsorgungstechnik
Postfach 3620
D-7500 Karlsruhe 1
W. Germany.

Gesellschaft für Strahlen- und Umwelt-Forschung mbH München
Institut für Tieflagerung
Theodor-Heuss-Strasse 4
D-3400 Braunschweig
W. Germany.

Sweden
Svensk Kärnbränslehantering AB
PO Box 5864
S-102 48 Stockholm
Sweden.

Swedish Geological Company
PO Box 1424
S-751 44 Uppsala
Sweden.

Switzerland
NAGRA
Parkstrasse 23
CH-5401 Baden
Switzerland.

United Kingdom
United Kingdom Atomic Energy Authority
Harwell, Oxfordshire
OX11 0RA
UK

British Geological Survey
Hicker Hill
Keyworth, Nottingham
NG12 5GG
UK

United States
National Technical Information Service
US Department of Commerce
5285 Port Royal Road
Springfield, Virginia 22161
USA

US Geological Survey
Books and Open-File Reports Section
Federal Center, Box 25425
Denver, Colorado 80225
USA

Superintendent of Documents
US Government Printing Office
Washington, DC 20402
USA

National Academy Press
National Academy of Sciences
2101 Constitution Avenue
Washington, DC 20418
USA

Index